D0200157

The Ape and the Sushi Master

Also by Frans de Waal

Chimpanzee Politics: Power and Sex Among Apes

Peacemaking Among Primates

*Good Natured: The Origins of Right and
Wrong in Humans and Other Animals*

Bonobo: The Forgotten Ape

The Ape
and the
Sushi Master

*Cultural Reflections
by a Primatologist*

Frans de Waal

BASIC
BOOKS

A Member of the Perseus Books Group

Copyright © 2001 by Frans de Waal

Published by Basic Books,
A Member of the Perseus Books Group

All rights reserved. Printed in the United States of America.
No part of this book may be reproduced in any manner
whatsoever without written permission except in the case of
brief quotations embodied in critical articles and reviews.
For information, address Basic Books, 387 Park Avenue South,
New York, NY 10016–8810.

Designed by Cynthia Young
Set in 11-point Electra by the Perseus Books Group.

Library of Congress Cataloging-in-Publication Data
Waal, F. B. M. de (Frans B. M.), 1948–
The ape and the sushi master : cultural reflections by a
primatologist / Frans de Waal.
p. cm.
Includes bibliographical references and index.
ISBN-13: 978-0-465-04176-3
ISBN-10: 0-465-04176-0
1. Psychology, Comparative.   2. Animal Behavior.
3. Human behavior.   I. Title.
QL785 .W126 2001
156—dc21
00-57922

First Paperback Edition

For Catherine: my rock

Contents

The Ape and the Sushi Master

Prologue

The Apes' Tea Party

Man is innately programmed in such a way that he needs a culture to complete him. Culture is not an alternative or replacement for instinct, but its outgrowth and supplement.

Mary Midgley, 1979

Jo Mendi, a cigar-smoking, brandy-drinking blue-collar worker with stocky legs and long arms, was used to having his name larger on the billboards than that of comedian Bob Hope, with whom he once co-starred. In the 1930s, he dominated Detroit's entertainment industry. Every day, he would show up in overalls at the side of the zoo director, who carried a cane and kept a watchful eye on his companion, who—being many times stronger than a grown man—had been known to molest unsuspecting bystanders. Such was Jo Mendi's fame

Jo Mendi showing his table manners with zoo director, John Millen, in the 1930s (Photo courtesy of the Detroit Zoological Society).

that the chimpanzee drew a crowd twice as large as the one that greeted the presidential candidate visiting the city, an issue Franklin Delano Roosevelt's opponents didn't hesitate to bring to the nation's attention.[1]

Petermann, a performing chimpanzee at the Cologne Zoo in the 1980s, was less lucky. Like Jo Mendi, he had a huge following and was, behind the scenes, not to be trifled with. His relationship with the zoo director was less amicable, however.

After attacking the director, Petermann was shot by the police. His fatal defiance of authority temporarily turned the ape into a martyr for the German anarchist movement.

Even today, Hollywood producers cannot resist throwing in a chimp or orangutan when their script asks for a laugh and they have failed to come up with anything better. An entire television show (*The Chimp Channel*) has been devoted to dressed-up apes trained to frantically move their mouths while an audio track with human speech gives the impression that they are talking.

The ape dinner parties that became standard at zoos and menageries in the nineteenth century, the chimpanzee entertainers of the twentieth century, and contemporary equivalents on the tube all project the image of animals doing their best to be like us, yet failing miserably. We get a kick out of such performances because our culture and dominant religion have tied human dignity and self-worth to our separation from nature and distinctness from other animals. Since we are the only ones who eat with cutlery—a sure sign of civilization—we are amused to see apes trying to do the same. They're not supposed to, and most certainly they're not supposed to be good at it. Lest the scene threaten the human ego, they must falter. As explained by Ramona and Desmond Morris:

> In the late 1920s the London Zoo started to organize these demonstrations on a regular basis. Each afternoon at a set time a group of young chimpanzees performed at a table

for the amusement of zoo visitors. They were trained to use bowls, plates, spoons, cups, and a tea pot. For the chimpanzee brain, learning to perform these trivial tasks provided only a minor challenge. There was the ever present danger that their table manners would become too polished. In order to relieve the monotony, it often became necessary to train them to "misbehave." They excelled at this, too, and their timing became so perfect that the tea cups were always popped into the teapot and the tea drunk through the spout, just at the vital moment when the keeper turned his back.[2]

To have apes ridicule our species, especially the cultural refinements that we admire so greatly in ourselves, could be looked at as a form of self-deprecation. That would be the optimistic view. The alternative is that by allowing animals to mock us we let them make even greater fools of themselves, which permits us to laugh away any doubts we might harbor about ourselves. That we select apes for this job is logical because it is particularly in the face of animals similar to us that human uniqueness needs confirmation.

To put this in perspective, imagine a family of elephants watching a television show in which people have hoses strapped to their noses and try to use the appendage to pick up a coin or uproot a small tree. The poor people in the show constantly get tangled up in their "trunks," trip over them, and in general demonstrate how ineptly unelephantine they are.

"They're getting restless. Okay boys, ham it up – knock the teapot over or something." (Cartoon of ape tea party at the zoo by Paul White, in 1962, reproduced with permission of Punch Ltd.)

I don't think we would find the show particularly funny, certainly not for longer than a couple of minutes, but an elephant family might never get enough of it.

This is because the issue is not humor, but self-definition.

Culture Versus Nature?

We define ourselves as the only cultured species, and we generally believe that culture has permitted us to break away from nature. We are wont to say that culture is what makes us human. The sight of apes with wigs and sunglasses acting as if they have made the same step is therefore utterly incongruous.

But what if apes have made this step to cultured behavior not only for the entertainment of the human masses, but also in real life without our assistance? What if they have their *own* culture rather than a superficially imposed human version? They might not be so amusing anymore. Indeed, even to contemplate such a possibility is bound to shake centuries-old convictions.

The possiblility that animals have culture is the topic I wish to explore in this book. Although such an exploration is worthwhile for a number of reasons, two stand out. First, there is growing evidence for animal culture—most of it hidden in field notes and technical reports—that deserves to be more widely known. Before we can consider this material, however, we need to temporarily abandon a few cherished connotations of the term "culture." This term evokes images of art and classical music, symbols and language, and a heritage that needs protection against the mass-consumption society. A so-called cultured person has achieved a refinement of tastes, a well-developed intellect, and a particular set of values and moral principles. This is not how scientists use "culture" in relation to animals. Culture simply means that knowledge and habits are *acquired* from others—often, but not always, the older generation—which explains why two groups of the same species may behave differently. Because culture implies learning from others, we need to rule out that each individual has acquired a particular trait by itself before we call it cultural.

The second reason for a book about animal culture is that it allows us to carry one more outdated Western dualism to its grave: the notion that human *culture* is the opposite of human *nature*. We in the West seem to have an uncontrollable urge to divide the world into two: good versus bad, us versus them, feminine versus masculine, learned versus innate, and so on. Dichotomies help organize our thinking, but they do so by neglecting complexities and shades of meaning. It is the rare thinker who keeps two contradictory thoughts simultaneously in mind; yet this is precisely what is often needed to get at the truth. Thus, while it is correct that learning affects all behavior, so does genetics, meaning that no behavior, whether human or animal, is dictated purely by one influence or the other.

In the last couple of years, the pendulum has swung away from nurture (or environmental effects) back to nature, leaving behind a number of bewildered social scientists who thought the issue had been settled. The current fascination with human biology, however, has created the opposite problem of people so obsessed with genetics that they ignore the other half of the equation. Twins-reared-apart studies have reached the status of common knowledge, and almost every week the media feature a new human gene. There is evidence for genes involved in schizophrenia, epilepsy, and Alzheimer's, and even in common behavioral traits such as thrill-seeking. Because genetic language ("a gene for *x*") plays into the hands of our sound-bite culture, we always need to add the warning that, by themselves, genes are like seeds dropped onto the pavement: in themselves they are powerless to produce anything. When scientists say that a trait is inherited, all they mean is that part of its variability is due to genetic factors. That the environment usually explains at least as much tends to be forgotten.

As Hans Kummer, a Swiss primatologist, remarked years ago, to try to determine how much of a trait is produced by genes and how much by the environment is as useless as asking whether the drum sounds that we hear in the distance are made by the percussionist or his instrument. On the other hand, if we pick up a *changed* drum sound, we can legitimately ask whether the difference is due to another drummer or another drum.[3] This is the only sort of question science addresses when it looks into genes versus the environment.

Culture is an environment that we create ourselves. For this reason, and quite contrary to the accepted view in some circles, culture does *not* deserve equal footing with nature. An entire generation of anthropologists has given this false impression by asking whether it is culture or nature that makes us act in a certain way. Natural selection, however, has produced our species, *including* our cultural abilities. Culture is part of human nature. To say that "man is made by culture," as many textbooks still do, is at the same level of accuracy as saying that "the river follows its bed." While true, the river also shapes its bed: the current river's flow is the product of the past river's action. In the same way, culture cannot exist apart from human nature, and there is profound circularity in saying that we are the product of culture if culture is the product of us.[4]

The relation between nature and culture reminds me of the mouse and the elephant walking side by side over a wooden bridge. Above the noise, the mouse shouts: "Hey, listen to us stamping together!" At the dawn of an undoubtedly Darwinian millennium, there are still those who claim that human behavior is mainly or entirely cultural. I see this exclusive focus like the mouse with delusions of grandeur walking next to human nature, the elephant who sets the tone of everything we do and are.

This is not to say that culture is mere icing on the cake, as some have suggested. Culture is an extremely powerful modifier—affecting everything we do and are, penetrating the core of human existence—but it can work only in conjunction with human nature. Culture takes human nature and bends it this way or that way, careful not to break it. That we have trouble looking through the false dichotomy is due to a peculiar uncertainty principle: we are unable to take off our cultural lenses, and hence can only guess at how the world would look without them. That is why we cannot discuss animal culture without seriously reflecting on our own culture and the possible blind spots it creates. Seemingly simple questions such as "Is there culture in nature?" and "Is there nature in culture?" cannot be answered without reflection on our own place in nature, a place that is culturally defined. I am not playing with words here. The only reason this sounds confusing is that we have been coaxed into treating nature and culture as opposites rather than closely intertwined.

Because of these larger issues, I find myself writing on topics on the margins of my expertise, from human goodness to Eastern philosophy, and from anthropomorphism to the aesthetic sense. Even if this is not the first time that I have stepped outside of my immediate field, which is to watch primates and prod them to give up their cognitive secrets, my task here is to discuss cultural biases, which makes me feel like a dog chasing its own tail, never really able to catch it. Moreover, one may question whether each culture fits in a little box: within cultures there is often plenty of disagreement. I often get the impression of being surrounded by two distinct categories of people: those who do and those who don't mind being compared with animals. I have encountered these contrasting attitudes among the great philosophers, among my teachers, and among friends and colleagues, and I have no idea what decides who will end up in which camp. It must have something to do with the empathy level towards animals, yet this just changes the question to why some people feel a connection with animals and others don't.

Around the World in Eighty Days

Part of my solution to the uncertainty principle has been to go places. If I cannot remove my cultural lenses, I can at least listen to people who grew up in other cultures. Thus, in the fall of 1998, I traveled around the world in eighty days. I went from Atlanta, where I live, to Austria, then China, Japan, Finland, and via the Netherlands, my native country, back to the United States. During this trip I researched several highly influential early students of human and animal behavior, such as the Austrian Konrad Lorenz, the Japanese Kinji Imanishi, and the Swedish-Finn Edward Westermarck. I debated monkeys and apes with colleagues in Japan, where cultural primatology found its origin. And most of all, I tried to weave together the three themes of this book: how we see other animals, how we see ourselves, and the nature of culture.

Each of these themes deserves a book of its own, but the special challenge of *The Ape and the Sushi Master* has been to freely move from one to the other, and from humans to other animals, while in the meantime poking a maximum number of holes in the nature/culture divide. In doing so, I have not striven for completeness, but picked issues that I felt best highlighted cultural biases in our treatment of nature, such as how noble or base we think our own species is, whether the bonobo's reception has been affected by how we judge its sexual morals, and how science goes about its business in the West and the East. These topics serve to illustrate how we selectively explore nature, sometimes shaping it in our own image.

My personal prejudices probably shine through, even though I may be less good at spotting them than some of my readers. I come from the southern part of the Netherlands. Since I was not born in the actual province of Holland, I rarely refer to my country by this name. The cruel hand of the Spanish Inquisition, which in the sixteenth century reached all the way to Flanders and my part of the Netherlands, put a halt to the Reformation that brought Calvinism to the North. The South stayed Roman Catholic, and as a result my up-bringing instilled less fear of God's wrath than is typical of the rest of Northern Europe. We have street carnivals (not unlike those in New Orleans), and in general we pride ourselves on a certain *joie de vivre*.

Like all Dutch children of my generation, I was taught German, French, and English in addition to my own lan-guage, and I learned these languages so well that I still speak them fluently. Fluency comes from practice, and even though I kept repeating as a child that I would never need all those stupid languages (I was more interested in science and math), later in life I married a French woman, moved to the United States, and began teaching and writing in English. It is hard to imagine anyone who has taken better advantage of his early language education!

The beauty of language is that each language is filled with concepts and expressions that reflect a distinct cultural out-look. Naturally, we try to translate these terms, but somehow the real flavor is only caught in the correct cultural and lin-guistic context. Words such as "date" or "cheerleader" may seem simple and obvious to Americans, but the concept of the-person-I-am-currently-going-out-with is unique to its cul-ture (not to mention the idea of a "blind date"), and the mystique surrounding cheerleading baffles every non-American. Similarly, French has a richer vocabulary for dishes, their taste, and their preparation than most non-Francophone people can even imagine, and the language brims with food-related expressions (such as "A kiss without moustache is like a soup without salt"). Every language captures a distinct way of looking at life, and no culture can ever be fully appreciated without an effort to speak the language.

As a European in America, one who crosses the Atlantic multiple times per year, I am very sensitive to the shades of value and meaning that go into making us the most cultural of cultural beings. I re-enter my Dutch persona when I am with my family, feel sort-of-French when visiting my in-laws in the Loire Valley, and after two decades in the upper Midwest and South, I am of course extensively familiar with the values, lifestyles, and cultural mix of the United States. So, even though this book may on occasion sound like an insult to the products of human culture, comparing them to the twigs and branches chimpanzees use in the jungle, I by no means want to trivialize how far we have come. I do wish to make the point, though, that culture must have had simple beginnings, some of which are to be found outside of our species.

In doing so, I will cover ground that one might think is safe for a scientist of my training, namely the question of animal culture. The ground is like quicksand, however. There is, in fact, so much resistance to the idea of animal culture that one cannot escape the impression that it is an idea whose time has come. The air is filled with claims and counterclaims; everyone has an opinion, and a strong one at that. In this melee, an entirely new field has come into existence: armchair primatology. It is not unusual for scholars barely able to tell a chimpanzee's front from its behind to criticize experts who have studied the species all their lives, or for someone who has never set foot on a particular island to dispute the findings of a team that has worked there for half a century. It must be a sign of its arrival that primatology has become everybody's business!

The island in question is Koshima, in the extreme south of Japan, where the first evidence for animal culture was gathered. A high point of my trip was a visit to Koshima, where I talked with the elderly but still sharp Mrs. Mito, who has been there from day one. After having heard so much about it, I was delighted to see with my own eyes how the monkeys still wash sweet potatoes in the ocean.

Litter-Box Culture

To introduce the topic of animal culture, let me start with an everyday example: the way cats learn to use the litter box. One of our cats visits the box to take a pee while her three kittens follow. Cats lack a sense of privacy, and so the offspring closely watch mom's activities, leaning over the rim of the box. Young kittens don't have particularly good eyesight, so it is unclear what they see. One of them awkwardly climbs into the box and is soon scratching around, moving litter like her mother. Then all of a sudden the little one, too, crouches and pees, with ears folded back. No one told her to do so, and it is hard to believe that cats have an inborn image of a modern invention such as the litter box. The mother's behavior seems to have triggered the youngster's.[5]

Social learning is widespread in animals and may continue well beyond the point at which it started. For example, as a student I worked in a laboratory in Utrecht where one scientist regularly caught monkeys out of a large group with a net. At first, the monkeys gave warning calls whenever they saw him approach with his dreadful net, but later they also did so when he only walked by. Still later, years after his research had ceased, I noticed that monkeys too young to have known the threat he once posed alarm-called for this man, and for no one else. They must have deduced from the reaction of their elders that he was not to be trusted. I recently heard that the group kept this alarm-call tradition up for decades, still always aimed at the same person!

The passing on of a "predator image" has also been observed in the field, in Kenya, by Dorothy Cheney and Robert Seyfarth. Vervet monkeys have different alarm calls for different predators (such as leopard, eagle, and snake), but need to learn to connect the one with the other. The investigators tested the knowledge of their monkeys by playing alarm calls from a concealed speaker. Since different predators require different responses, they divided the reactions of infant monkeys into three types: one was to run to the safety of mother, another was to react in a way that could get them into trouble, and the third was the correct response. For example, the right response to a snake alarm is to stand upright in the grass and look around. This would be suicidal in reaction to a leopard call, which requires monkeys to climb a tree. Cheney and Seyfarth found that both mother-oriented and wrong responses disappeared with age, whereas correct responses increased. This suggests that young monkeys learn how to react to each specific alarm. It would be incredibly costly for them to do so by trial and error: most likely they pick up information from the rest of the group. Indeed, youngsters who watched adults before responding themselves were more likely to show the right response to a call.[6]

These findings contradict the widespread belief that survival tactics must be hard-wired and instinctive. Not so in the case of vervet alarm calls: monkeys who fail to pay attention to their fellows simply will not make it. What we have, then, is an absolutely critical set of responses transmitted through the observation of others. Instead of relying on genetic information, this is a social, cultural process. In the laboratory, Susan Mineka has demonstrated the same kind of learning by showing snakes to captive-born monkeys who had never seen one in their entire lives. These naïve monkeys were unafraid until they saw their wild-born parents react with intense fear to the same snakes. From that day on, the captive-born monkeys, too, showed snake phobia.[7] And not only in primates do we see a cultural construction of the enemy image: the same has been demonstrated in birds.[8]

The other major domain of cultural learning is food. Animals learn from each other what to eat, and what not. Parent crows that fly daily with their offspring to the local garbage dump to look for tasty morsels instill in them a life-long preference for such sites, whereas the crow family that survives on natural foods will have offspring that carry on the same tradition when they get older. Food aversion is similarly transmitted. This was first noticed by a German rodent-control officer who set out poisoned bait, killing wild rats in large numbers. After a while, however, the remaining rats began to avoid the bait, and their offspring would do the same. Without any direct experience with the bait, young rats would eat only safe foods.

An experimental psychologist, Bennett Galef, tested this in his laboratory by feeding rats two diets of different texture, taste, and smell. He then laced one of the diets with lithium chloride, which makes rats sick. This procedure led the animals to avoid the contaminated diet. The question now was how the rats' offspring would react after removal of the contamination. Both diets were again perfectly okay to eat, but adults fed exclusively on only one diet due to their bad experience with the other. It turned out that the pups acted like their parents. Of 240 pups given a choice of both diets, only one ate any of the food that adults in its colony had learned to avoid.[9]

All of these examples—the alarm calling, the snake fear, the food aversion—arouse intense debate among psychologists about the exact learning mechanism involved. One might think it is mere imitation, but increasingly the term "imitation" is being reserved for cases in which a solution to a problem is copied with an *understanding* of both the problem and the model's intentions. This usage has turned "imitation" into a small, cream-of-the-crop subset of social learning, one that may not apply to rats and cats, perhaps not even to monkeys and apes.

As soon as individual learning enters the picture—that is, a behavior is acquired partially through trial and error—the suspicion is that we are dealing with something simpler than imitation. A good example is the kittens and the litter box: it is very possible that all they learn from their mother is *where* to do the deed. Once they have been brought to the right spot, the rest can be construed as regular feline responses to the smell of urine and the feeling of loose gravel under their feet. So, even though the kittens act like Mom, this doesn't necessarily mean that they are following her example, and even less that they understand the box's purpose.

To dismiss such behavior as having nothing to do with imitation is not altogether fair to animals, though, because we don't apply the same standards to people. If, from watching a soccer game, I develop the habit of kicking a ball, this doesn't mean that all there is to becoming a player is to mimic observed actions. It takes years of practice to control the ball and send it whichever way I want it to go. If one could become a soccer star from merely watching the game played by others,

the world would be full of them. All imitation is a combination of a general idea picked up from others and individual
practice to refine the skill. Inasmuch as we accept this simple
truth for human imitation, why be so picky about other animals? True, they often have only a vague understanding of
what others are doing—if they understand it at all—but whatever information they gain from watching is built into a solution developed by themselves, which is the way we imitate
much of the time as well.

The simplest form of social learning is known as "local enhancement," in which one individual is attracted to a place
where another is doing something interesting, such as finding
food. The attraction then leads the first individual to explore
the same situation and learn the solution on its own. The
model thus indicates the *where* rather than the *how* of the answer. Our kitten example fits this category.

Another common possibility is known in anthropology as
"stimulus diffusion" and in psychology as "emulation." Here a
general idea, outcome, or concept is obtained from others but
the specifics are worked out independently. A modern-day example is how Microsoft "borrowed" the windows concept
from Macintosh. DOS-based machines now produce approximately the same clickable environment on the screen as the
Macintosh, but they do so via a totally different programming
architecture. Microsoft thus rightly claims that Windows is
not an Apple imitation: it is a mere emulation. Similarly, a
bird may learn from another that crabs can be opened and

that the inside is edible, but will still need to figure out on its own how to get to these softer parts.

Whatever the exact process, the critical question before we speak of culture is whether an animal would ever have hit on a particular solution or developed a particular habit without the benefit of social companions. Would my kittens have learned to use the litter box on their own? I am afraid not. Would the captive-born monkeys have come to fear snakes on their own? Yes, but only after having been bitten, which is a far trickier way of getting to know snakes than through the observation of others. Social learning has tremendous advantages. We can debate long and hard what to call the process or how complex it is. All that really matters is that one individual adopts a habit under the influence of another.

The Sushi Master

Learning from others is second nature to humans: we do it more readily and precisely than any other animal. Therefore, when a young chimpanzee is raised with a human child, the direction of influence is more likely to be from the ape to the child than the other way around. This was discovered the hard way, in the 1930s, by Winthrop and Luella Kellogg, who were forced to terminate a co-rearing experiment in their home when their son, Donald, began to give guttural food barks like those of the female chimpanzee, Gua, with whom he was being raised. When Donald picked up an orange and ran to his parents while grunting "uhuh, uhuh," it was decided that his aping of the ape had gone far enough:

> The situation in which the two lived together as playmates and associates was much like that of the two-child family in which Gua, because of her greater maturity and agility, played the part of the older child. With the added stimulation thus afforded, the younger child in such situations usually learns more rapidly than would otherwise be the case. It was Gua, in fact, who was almost always the aggressor or leader in finding new toys to play with and new methods of play; while the human was inclined to take up the role of the imitator and follower.[10]

Gua, too, was a good imitator. The Kelloggs describe how she became a typist after having seen her foster parents type for many months. One day, a very young Gua climbed on the typewriter stool and sat properly behind the machine, moving her hands simultaneously up and down the keyboard, pounding the keys with her fingers. We can only speculate about the literary heights the chimpanzee might have attained had the experiment continued.

There are now many studies of the mimetic abilities of apes, such as the one by Deborah Custance at the Yerkes Primate Center in Atlanta. Sitting in front of two juvenile chimpanzees, Scott and Katrina, the investigator would make simple gestures, such as raising a foot, slapping the floor, or wiping her own face, and reward the apes for copying them. After this training, Custance demonstrated a series of gestures that had never been rewarded before. These included puffing out her cheeks, clapping, jumping, and self-hugging. Scott and Katrina's responses were videotaped and evaluated by observers kept in the dark about what the experimenter had demonstrated. This way, there was an independent assessment of the imitation. The two apes did very well, showing that they had no trouble copying arbitrary body movements.[11] We may not realize it—being ourselves masters of imitation—but the translation of perceived into performed action is quite a feat. The tendency to act like behavioral Xerox machines sets apes apart from most other animals and makes them obvious candidates for the evolution of culture.

Masako Myowa-Yamakoshi and Tetsuro Matsuzawa, at the Primate Research Institute of Kyoto University, conducted a more complex study of imitation involving all kinds of objects. Matsuzawa runs a facility in which chimpanzees live outdoors in a social group but can be called inside for a voluntary experiment. Once the ape sits in front of him, the human demonstrates a simple action. All of the apes in this study were fully adult and hence probably less inclined to imitate than juveniles. Seeing each action only once, the apes rarely copied them. They did so only if the action linked two objects (such as putting a ball in a bowl) rather than linking an object with the body (such as putting a bowl on one's head). Interestingly, connecting different objects is typical of tool use in the field, such as when chimpanzees poke a stick into a termite hill or use chewed leaves as a sponge to extract water from a hole. Could it be that the ape mind is set up to pay special attention to technical solutions so as to better replicate them?[12]

Under normal circumstances, apes see the behavior of their group mates numerous times, and so have many opportunities to become familiar with them. They watch others at close range, following each and every move in detail. Perhaps, as suggested by Matsuzawa, they follow the model of the sushi-master apprentice. The apprentice slaves in the shadow of masters of an art requiring rice of the right stickiness, delicately cut ingredients, and the simple, eye-catching arrangements for which Japanese cuisine is known. Anyone who has tried—as have I—to cook rice, mix it with vinegar, and cool it off with a hand-held fan so as to quickly mold fresh rice balls in one's hands, knows what an incredibly complex skill this is, and it is only a small part of the job. Actually, I have been told that the reason one never sees female sushi masters is that a woman's hands are too warm for the task—an explanation to be taken with a grain of salt, given that no one ever complains about the sushi that women prepare at home. Men tend to claim high-status jobs for themselves; the exclusion of women from the sushi domain confirms its central place in Japanese culture.

To return to the apprentice sushi master: his education seems a matter of passive observation. The young man cleans the dishes, mops the kitchen floor, bows to the clients, fetches ingredients, and in the meantime follows from the corners of his eyes, without ever asking a question, everything that the sushi masters are doing. For no less than three years he watches them without being allowed to make actual sushi for the patrons of the restaurant—an extreme case of exposure without practice. He is waiting for the day on which he will be invited to make his first sushi, which he will do with remarkable dexterity.

This runs counter to imitation the way I described it before, in which an idea caught from others is supplemented with a great deal of individual practice. But who knows what apprentices do in their spare time? It is entirely possible, for instance, that the older masters—who, like all aging male primates, are more patient with younger males—take the apprentice aside after the closing of the restaurant to show him a few tricks and have him try things out for himself. Whatever the truth about the sushi master's education, Matsuzawa's point is that the watching of skilled models firmly plants action sequences in the head that come in handy, sometimes much later, when the same task needs to be carried out.

It Takes a Village ...

Watching others is a favorite activity of young primates. They constantly hang around their elders, absorbing every little detail of what is going on. At the same time that psychologists are debating what young animals do with all of this information, and whether it deserves to be called imitation if they duplicate the actions of others, field-workers have taken an entirely different tack to the issue of animal culture. Much like cultural anthropologists, who document how one human population differs from another, they compare different sites and note how each chimpanzee community has its own way of doing things. This ethnographic method is also being applied to other animals, most successfully to dolphins and whales. The rapidly growing literature gives the impression that we have only scratched the surface: cultural diversity in the animal ✳ kingdom probably takes on vast proportions.

Whereas these observations are not being contested, not everyone agrees that the term "culture" best describes differences between groups. This obviously depends on one's definition. You would think that scholars dispassionately arrive at a reasonable characterization of a phenomenon, after which they only need to agree on what is and is not covered by it. But definitions are rarely neutral; they mirror entire world views. Behind the ongoing culture wars, the debate is about nothing less than humanity's place in the cosmos. Definitions of culture have become the political football in this larger controversy.

It is not hard to come up with a definition of culture that rules out all species except our own. Even tools can be defined in such a way that they are found only in our species—for example, by requiring that they fit a symbolic context. Such exclusive definitions tend to focus on the highest human achievements associated with a process, declaring these absolutely essential. This is a legitimate line of thought, inasmuch as it lets scientists comfortably speak of the uniquely human capacities for culture, tool use, language, morality, and politics.

My own bias, however, and that of many fellow primatologists, is quite the opposite. We tend to look beyond the brief evolutionary history of the human race, eyeing a much longer past and a much wider range of animals. All the fancy things that humans do with tools and culture are certainly worthy of attention, but they are best kept out of initial definitions so as to cast the net as widely as possible. This approach is commonplace in biology. Thus, biologists are comfortable saying that both chickens and people walk bipedally even though it is obvious that they do so in radically different ways (look at which way their "knees" are pointing!). Biologists always define processes—nutrition, locomotion, reproduction—in the broadest possible terms because evolution has produced a multitude of means to achieve them.

Broad definitions have the additional advantage that they permit us to see the full range of a phenomenon. For example, one could define language so narrowly that the babbling of a toddler does not fall under it, but does this mean that babbling has nothing to do with language? Narrow definitions neglect boundary phenomena and precursors, and they often mistake the tip of the iceberg for the whole. Thus, by saying, as some have done, that in the absence of teaching and instruction there is no point in speaking of culture, one immediately throws out a multitude of human cultural traits. Many habits are picked up without any instruction whatsoever: they require mere exposure, day in day out, to a particular cultural context. The warmth or spontaneity with which we treat our fellow citizens, the way our taste buds react to spices, the desire for consensus over confrontation, the melody and loudness of our voices—all of these become so ingrained that we call them "second nature." They are profoundly cultural, however, despite the fact that active teaching has very little to do with their establishment.

For the biologist, the way habits are transmitted is secondary. All we care about is whether the process is "visible" to natural selection. That is, does learning from others contribute to survival? As illustrated by the examples of alarm calling, food aversion, and snake fear, there is every reason to believe that, yes, information gained from others plays a major role in the struggle for existence. Rehabilitation programs, in which home-reared apes have been released into the wild, have taught us how critical it is for these animals to know what to eat, where to go, and what to avoid. Having grown up in the absence of adult models of their species, young apes are rarely successful in the forest, often starving to death. In this sense, apes are as culturally reliant as we are.

The same is true for vulture culture, and not just because it rhymes. When the last few remaining wild California condors were rounded up in the 1980s to establish an artificial breed-

ing program at zoos, it was decided to feed the first generation
of chicks with hand puppets in the color and shape of adult
members of their species. The idea was that this was all that
was needed to turn them into real condors. Despite the pup-
pet show, however, the young condors learned to associate humans with food. Upon release into the wild, they ended up
hanging around human dwellings, incapable of scavenging on
their own. The normally shy, magnificent foragers had been
turned into barnyard chicks perching on rooftops. Evidently,
the hand-reared vulture is as culturally disadvantaged as the
hand-reared ape.

The standard notion of humanity as the only form of life to
have made the step from the natural to the cultural realm—as
if one day we opened a door to a brand-new life—is in urgent
need of correction. The transition to culture has no doubt
been gradual, in small incremental steps, and was neither complete (we never left human nature behind) nor much different,
at least initially, from the behavioral traditions seen in other animals. The idea that we are the only species whose survival depends on culture is false, and the entire endeavor of juxtapos-
ing nature and culture rests on a giant misunderstanding.

In *Consilience* (1998), Edward Wilson offered a Darwinian
embrace to the social sciences. Some academics no doubt experienced his gesture as suffocating, but it cannot be denied
that increased integration among the behavioral sciences is
sorely needed. Wilson did extensively discuss the same na-
ture/culture divide that is at issue here, but the interdiscipli-

nary bridge that he tried to build started at the other end. Instead of urging the social sciences and humanities to absorb more biology, I am asking them to carefully reconsider their own chosen domain—often defined in opposition to biology—and see how broadly it applies. They can export their ideas to students of animal behavior, who will agree that the social environment directs development, and that each individual is part of a larger whole in both body and mind: the group, troop, colony, flock, or community. Imagine that the African proverb "It takes a village to raise a child" applies to baboons, elephants, or dolphins: an entirely new perspective on the social life of animals will ensue. This perspective would be quite close to that of the social sciences, drawing on ways of thinking now applied uniquely to our own species.

At the same time, there is no doubt that we have taken culture an unprecedented step farther than other animals because of symbols, language, ideas, meanings, values, teaching, and imitation. In this sense, the human cultural capacity is truly unique, and has become so pervasive in our lives that it is no surprise that we marvel at its power. Not only do we create cultures, but once created they lend meaning and feed back into everything we do, transforming the very core of our being. We both produce and are produced by culture to a degree not found in any other animal.

Perhaps this is due to our ability, stressed by Michael Tomasello, to build new inventions upon older ones. Tomasello calls the accumulation of improvements through history a "ratchet effect," which he sees as uniquely human.[13] I have some qualms, because there seems no intrinsic reason why knowledge accumulation would be hard for animals. It seems unlikely that complex sequences of coordinated actions, such as nut cracking by chimpanzees or beach hunting by killer whales, were invented all at once: what we see these animals do today is most likely the endpoint of a long and steady perfection of skills.[14] On the other hand, even if other animals occasionally elaborate upon previous achievements, there can be no doubt that they do so on a smaller scale than we do. Any such difference would be greatly magnified over multiple generations. Possibly, then, the ratchet effect is the yeast in the dough of human culture.

But despite our cultural superiority, what harm can there be in exploring nonhuman parallels to human cultural capacities? Are we only happy with a day-and-night difference, in which we have it all and other animals nothing? Imagine that we were to define "eating" by the use of knife and fork. Such a definition would allow us to claim eating as uniquely human, even uniquely Western, yet we would accomplish this distinction by confusing the instruments of consumption with its essence. The essence of eating is to get food into one's stomach, and in this regard we are obviously not special at all. The relevant question in relation to culture, therefore, is, what is its essence? What is the least common denominator of all things called cultural? In my view, this can only be the nongenetic spreading of habits and information. The rest is nothing else than embellishment. Those who have elevated language, education, values, and other typically human aspects of culture to its defining criteria confuse the knives and forks of the process with its essence. In doing so, they have succeeded in keeping other animals out at the expense of a larger picture, one with the potential of revealing a glimpse of our own cultural origins.

My own definition of culture reflects this broader view:

> *Culture is a way of life shared by the members of one group but not necessarily with the members of other groups of the same species. It covers knowledge, habits, and skills, including underlying tendencies and preferences, derived from exposure to and learning from others. Whenever systematic variation in knowledge, habits, and skills between groups cannot be attributed to genetic or ecological factors, it is probably cultural. The way individuals learn from each other is secondary, but that they learn from each other is a requirement. Thus, the "culture" label does not apply to knowledge, habits, or skills that individuals readily acquire on their own.*

If history has taught us anything, it is to be cautious in postulating differences. It is not too long ago that it was said that "savages" were incapable of organizing themselves into societies, that the word "society" really didn't apply to people marked by rampant promiscuity, crime, and laughably simple languages. Now we realize, of course, that all humans, including those in preliterate societies, have complex value systems and moral rules, and that they speak languages every bit as rich as the one you are now reading.

There exists a parallel history of misconceptions about our primate relatives, who entered Western thinking as the incarnation of the devil, put on earth to mock the crown of creation. These animals have been underestimated over and over, and erosion of common misconceptions has been slow. Whenever their abilities are said to approach ours, the reaction is often furious. For example, claims of language abilities in apes became so threatening that at one international conference, in 1980, there was an unsuccessful move to *ban* all animal language research, similar to a ban on the study of language origins by the Linguistic Society of Paris in 1866.[15] I am not saying that apes are capable of language, but attempts at censorship do reveal just how much insecurity surrounds human uniqueness.

No wonder that the idea of animal culture had to come from the East, where human self-definition doesn't hinge on a Freudian defeat of basic impulses or a denial of the connection with nature. Given how much the culture concept is tied to the idea that we have distanced ourselves from other animals, this book must explore how animal-like we are, or how humanlike animals are. It must also return to such classical clashes—still as relevant now as then—as those between behaviorists and ethologists, who emphasized learning and instinct, respectively. At every turn I will try to undermine existing dualisms, always looking for the more integrated picture.

In the meantime, it is evident that we have lost control over the apes' tea party. Instead of imitating us, and knocking over the teapot at our expressed request, the apes have taken over the show, displaying habits that they themselves developed and tricks we didn't teach them. As a result, they're holding an entirely different mirror up to us, one in which apes are not human caricatures but serious members of our extended family with their own resourcefulness and dignity.

Ever since Carl Linnaeus courageously classified us with the monkeys and apes in 1758, the message has been coming at us that we are not alone. Biologically speaking, we never were. The time has come to argue the same with regard to culture.

Cultural Glasses

The Way We See Other Animals

The Western world's historic lack of exposure to monkeys and apes has only reinforced its sense of human uniqueness. Ever since Descartes, the air has been filled with warnings against anthropomorphism. The charge is that we love to project thoughts and feelings onto animals, making them more humanlike than they are.

But getting rid of anthropomorphism is neither easy nor risk-free. By changing our language as soon as we describe animals, we may be concealing genuine similarities. When pioneers of the naturalistic study of animal behavior began to emphasize continuities with human behavior in the 1960s, the message was shocking. It has been amplified since by the burgeoning fields, from primatology to sociobiology, that they helped spawn.

Even a quintessentially human activity, such as art, has not been exempted from such claims. Given that our aesthetic sense has been shaped by the environment in which we evolved, it is logical to expect preferences for shapes, contrasts, and colors to transcend species. Hence we should not be surprised if a composer as great as Mozart admired one as small as his pet starling.

The Whole Animal

Childhood Talismans and Excessive Fear of Anthropomorphism

"Why do I tell you this little boy's story of medusas, rays, and sea monsters, nearly sixty years after the fact? Because it illustrates, I think, how a naturalist is created. A child comes to the edge of deep water with a mind prepared for wonder. He is given a compelling image that will serve in later life as a talisman, transmitting a powerful energy that directs the growth of experience and knowledge."

Edward O. Wilson, 1995

"Fear of the dangers of anthropomorphism has caused ethologists to neglect many interesting phenomena, and it has become apparent that they could afford a little disciplined indulgence."

Robert Hinde, 1982

Scientists are supposed to study animals in a totally objective fashion, similar to the way we inspect a rock or measure the circumference of a tree trunk. Emotions are not to interfere with the assessment. The animal-rights movement capitalizes on this perception, depicting scientists as devoid of compassion.

Some scientists have proudly broken with the mold. Roger Fouts, known for his work with language-trained chimpanzees, says in *Next of Kin:* "I had to break the first commandment of the behavioral sciences: *Thou shalt not love thy research subject.*" Similarly, Jeffrey Masson and Susan McCarthy, in *When Elephants Weep,* make it seem that very few scientists appreciate the emotional lives of animals.

In reality, the image of the unloving and unfeeling scientist is a caricature, a straw man erected by those wishing to pat themselves on the back for having their hearts in the right place. Unfeeling scientists do exist, but the majority take great pleasure in their animals. If one reads the books of Konrad Lorenz, Robert Yerkes, Bernd Heinrich, Ken Norris, Jane Goodall, Cynthia Moss, Edward Wilson, and so on, it becomes impossible to maintain that animals are invariably studied with a cold, callous eye.

I have met many other scientists who may not write in the same popular style—and who may not dwell on their feelings, considering them irrelevant to their research—but for whom the frogs, budgerigars, cichlid fish, bats, or whatever animals they specialize in hold a deep attraction. How could it be otherwise? Can you really imagine a scientist going out every day to capture and mark wild prairie voles—getting bitten by the voles, stung by insects, drenched by rain—without some deeper motivation than the pursuit of scientific truth? Think of what it takes to study penguins on the pack ice of the Antarctic, or bonobos in hot and humid jungles overrun by armed rebels. Equally, researchers who study animals in captivity really need to like what they are doing. Care of their subjects is a round-the-clock business, and animals smell and produce waste—which some of my favorite animals don't mind hurling at you—something most of us hardly think about until we get visitors who hold their noses and try to escape as fast as they can.

I would turn the stereotype of the unfeeling scientist around and say that it is the rare investigator who is not at some level attached to the furry, feathered, or slippery creatures he or she works with. The maestro of observation, Konrad Lorenz, didn't believe one could effectively investigate an animal that one didn't love. Because our intuitive understanding of animals is based on human emotions and a sense of connection with animals, he wrote in *The Foundations of Ethology* (1981) that understanding seems quite separate from the methodology of the natural sciences. To marry intuitive insight with systematic data collection is both the challenge and the joy of the study of animal behavior.

Attraction to animals makes us forget the time spent watching them, and it sensitizes us to the tiniest details of behavior. The scientific mind uses the information thus gathered to formulate penetrating questions that lead to more precise research. But let us not forget that things did not start out with a scientific interest: the lifeblood of our science is a fascination with nature. This always comes first, usually early in life. Thus, Wilson's career as a naturalist began in Alabama, where as a boy—in an apparent attempt to show that not all human behavior is adaptive—he used his bare hands to pull poisonous snakes from the water. Lorenz opened his autobiographical notes for the Nobel Committee with "I consider early childhood events as most essential to a man's scientific and philosophical development." And Goodall first realized that she was born to watch animals when, at the age of five, she entered a chicken coop in the English countryside to find out how eggs were made.

Closeness to animals creates the desire to understand them, and not just a little piece of them, but the *whole* animal. It makes us wonder what goes on in their heads even though we fully realize that the answer can only be approximated. We employ all available weapons in this endeavor, including extrapolations from human behavior. Consequently, anthropomorphism is not only inevitable, it is a powerful tool. As summed up by Italian philosopher Emanuela Cenami Spada:

> Anthropomorphism is a risk we must run, because we must refer to our own human experience in order to formulate questions about animal experience. . . . The only available "cure" is the continuous critique of our working definitions in order to provide more adequate answers to

our questions, and to that embarrassing problem that animals present to us.[16]

The "embarrassing problem" hinted at is, of course, that we see ourselves as distinct from other animals yet cannot deny the abundant similarities. There are basically two solutions to this problem. One is to downplay the similarities, saying that they are superficial or present only in our imagination. The second solution is to assume that similarities, especially among related species, are profound, reflecting a shared evolutionary past. According to the first position, anthropomorphism is to be avoided at all cost, whereas the second position sees anthropomorphism as a logical starting point when it comes to animals as close to us as apes.

Being a proponent of the second position creates a dilemma for an empiricist such as myself. I am not at all attracted to cheap projections onto animals, of the sort that people indulge who see cats as having shame (a very complex emotion), horses as taking pride in their performance, or gorillas as contemplating the afterlife. My first reaction is to ask for observables: things that can be measured. In this sense, I am a cold, skeptical scientist. With my team of students and technicians, I watch primates for hundreds of hours before a study is completed, entering codes of observed behavior into handheld computers. We also conduct experiments in which chimpanzees handle joysticks to select solutions to problems on a computer screen. Or we have monkeys operate an apparatus that allows them to pull food toward themselves, after which we see how willing they are to share the rewards with those who assisted them.[17]

All of this research serves to produce evidence for or against certain assumptions. At the same time that I am committed to data collection, however, I argue for breathing space in relation to cognitive interpretations, don't mind drawing comparisons with human behavior, and wonder how and why anthropomorphism got such a bad name. Anthropomorphism has proven its value in the service of good, solid science. The widely applied vocabulary of animal behavior, such as "aggression," "fear," "dominance," "courtship," "play," "alarm," and "bonding," has been borrowed straight from language intended for human behavior. It is doubtful that scientists from outer space, with no shared background to guide their thinking, would ever have come up with such a rich and useful array of concepts to understand animals. To recognize these functional categories is the part of our job that comes without training and usually builds upon long-standing familiarity with pets, farm animals, birds, bugs, and other creatures.

In my own case it began with a love for aquatic life.

Zigzag through the Polder

Almost every Saturday when I was a boy, I jumped on my bike to go to the polder, a Dutch word for low-lying land reclaimed from the water. Bordering the Maas River, our polder was dissected by freshwater ditches full of salamanders, frogs, stickleback fish, young eels, and water insects. Carrying a crudely constructed net—a charcoal sieve attached to a broomstick—I would jump over ditches, occasionally sliding into them, to get to the best spots to catch what I wanted. I returned in a perilous zigzag, balancing a heavy bucket of water and animals in one hand while steering my bike with the other. Back home, I would release my booty in glass containers and tanks, adding plants and food, such as water fleas caught with a net made out of one of my mother's old stockings.

Initially, the mortality in my little underwater worlds was nothing to brag about. I learned only gradually that salamanders don't eat things that don't move, that big fish shouldn't be kept with little ones, and that overfeeding does more harm than good. I also became aware of the ferocious, sneaky predation by dragonfly larvae. My animals started to live longer. Then one day—I must have been around twelve—I noticed a dramatic color change in one of my sticklebacks in a neglected tank with unchecked algae growth. Within days, the fish turned from silvery to sky blue with a fiery red underbelly. A plain little fish had metamorphosed into a dazzling peacock! I was astonished and spent every free minute staring into the aquarium, which I didn't clean on the assumption that perhaps the fish liked it better that way.

This is how I first saw the famous courtship behavior of the three-spined stickleback. The two females in the tank grew heavy bellies full of roe, while the male built a nest out of plant material in the sand. He repeatedly interrupted his hard work by performing a little dance aimed at the females, which took place closer to the nest site each time. I did not understand everything that was going on, but I did notice that the females suddenly lost their eggs, whereupon the male started moving his fins rapidly (I later learned that his fanning served to create a current to send additional oxygen over the eggs). I ended up with a tank full of fry. It was an exhilarating experience, but one that I had to enjoy all by myself. Although my family tolerated my interests, they simply could not get excited about a bunch of tiny fish in one of my tanks.

I had a similar experience years later, when I was a biology student at the University of Nijmegen. In a welcome departure from the usual emphasis on physiology and molecular biology, one professor gave a lecture on ethology—the naturalistic study of animal behavior—featuring detailed drawings of the so-called zigzag dance of the stickleback. Because of the work of Niko Tinbergen, a Dutch zoologist, the stickleback's display had become a textbook example. The drawings of my professor were wonderful, showing the male pushing out his red belly, with spines pointing outward, then leading the female to the nest while performing abrupt back-and-forth movements in front of her. When I nudged my fellow students, excitedly telling them that I knew all this, that anyone could see it in a small aquarium at home, once again I met with blank stares. Why should they believe me, and what was the big deal about fish behavior, anyway? Didn't I know the future was in biochemistry?

A few years later, Tinbergen received a Nobel Prize: the stickleback had won! By that time, however, I had already moved to Groningen, a university where ethology was taken more seriously. I now study the behavior of monkeys and apes. This may seem incongruent given my early interests, but I have never had a fixation on a particular animal group. There simply weren't too many chimpanzees in the polder; otherwise I would have brought them home as well.

One thing bothered me as a student. In the 1960s, human behavior was totally off limits for the biologist. There was animal behavior, then there was a long time nothing, after which came human behavior as a totally separate category best left to a different group of scientists. This way we kept the peace, because the other scientists were—to borrow a concept from animal behavior—pretty territorial. Popular books by Desmond Morris (*The Naked Ape*) and Lorenz (*On Aggression*) were extremely controversial because they voiced continuity between human and animal behavior. If young students of animal behavior now look down upon these authors, seeing themselves as far more sophisticated, they forget how much they owe them for knocking down the walls well before the sociobiological revolution came along. I wasn't able to judge the scientific merit of their work then, but something about these ethologists felt absolutely right: they saw humans as animals. It is only in reading them that I real-

ized that this was the way I had felt for as long as I could remember.

Pecking Orders in Oslo

It is hard to name a single discovery in animal behavior that
has had a greater impact and enjoys wider name recognition
than the "pecking order." Even if pecking is not exactly a human behavior, the term is ubiquitous in modern society. In
speaking of the corporate pecking order, or the pecking order
at the Vatican (with "primates" on top!), we acknowledge both
inequalities and their ancient origins. We also slightly mock
the structure, hinting that we, sophisticated human beings
that we are, share a few things with domestic fowl.

The momentous discovery of rank orders in nature was
made at the beginning of the twentieth century by a
Norwegian boy, Thorleif Schjelderup-Ebbe, who fell in love
with chickens at the tender age of six.[18] He was so enthralled
by these sociable birds that his mother bought him his own
flock at a rented house outside of Oslo. Soon each bird had a
name. By the age of ten, Thorleif was keeping detailed note-
books, which he maintained for many years. Apart from keeping track of how many eggs his chickens laid, and who pecked
whom, he was particularly interested in exceptions to the hierarchy, so called "triangles," in which hen A is master over B,
and B over C, but C over A. So, from the start, like a real scientist, he was interested in not only the regularities but also

the irregularities of the rank order. The social organization that he discovered is now so obvious to us that we cannot imagine how anyone could have missed it, but no one had described it before.

The rest is history, as they say, but not a particularly pretty one. The irony is that the discoverer of the pecking order was himself a henpecked man. Thorleif the boy had a very domineering mother, and later in life he ran into major trouble with the very first woman professor of Norway. She supported him initially, but as an anatomist she had no real interest in his work.

After Schjelderup-Ebbe received a degree in zoology, he published the chicken observations of his youth while coining the term *Hackordnung*, German for pecking order. His classic paper, which appeared in 1922, describes dominants as "despots" and demonstrates the elegance of hierarchical arrangements in which every individual has its place. Knowing the rank order among 12 hens, one knows the dominance relation in all 66 possible pairs of individuals. It is easy to see the incredible economy of description, and to understand the discoverer's obsession with triangles, which compromise this economy.

At about the time that the young zoologist wanted to continue his studies, however, a malicious but well-written piece in a student paper made fun of his professor. An enemy then spread the rumor that the anonymous piece had been written by Schjelderup-Ebbe, who was indeed a gifted writer. Even though the piece was actually written by Sigurd Hoel, later to become one of Norway's foremost novelists, irreparable damage had been done to the relationship with his professor. She withdrew all support and became an active foe. As a result of lifelong intrigues against him, Thorleif Schjelderup-Ebbe never obtained a Norwegian doctorate, and never received the recognition he deserved.

Regardless of this sad ending, the beginning of the story goes to show how a child who takes animals seriously, who considers them worthy of individual recognition, and who assumes that they are not randomly running around but, like us, lead orderly lives, can discover things that the greatest scientists have missed. This quality of the child, of unhesitatingly accepting kinship with animals, was remarked upon by Sigmund Freud:

> Children show no trace of the arrogance which urges adult civilized men to draw a hard-and-fast line between their own nature and that of all other animals. Children have no scruples over allowing animals to rank as their full equals. Uninhibited as they are in the avowal of their bodily needs, they no doubt feel themselves more akin to animals than to their elders, who may well be a puzzle to them.[19]

The intuitive connection children feel with animals can be a tremendous source of joy. The unconditional love received from pets, and the lack of artifice in the relationship, contrast sharply with the much trickier dealings with members of their own species. I had an animal friend like this when I was young; I still think fondly of the neighbors' big dog, who was often by my side, showing interest in everything I did or said. The child's closeness to animals is fed by adults with anthropomorphic animal stories, fairy tales, and animated movies. Thus, a bond is fostered with all living things that is critically examined only later in life. As explained by the late Paul Shepard, who like no one else reflected on humanity's place in nature:

> Especially at the end of puberty, the end of innocence, we begin a lifelong work of differentiating ourselves from them [animals]. But this grows from an earlier, unbreakable foundation of contiguity. Alternatively, a rigorous insistence of ourselves simply as different denies the shared underpinnings and destroys a deeper sense of cohesion that sustains our sanity and keeps our world from disintegrating. Anthropomorphism binds our continuity with the rest of the natural world. It generates our desire to identify with them and learn their natural history, even though it is motivated by a fantasy that they are no different from ourselves.[20]

In this last sentence, Shepard hints at a more mature anthropomorphism in which the human viewpoint is replaced, however imperfectly, by the animal's. As we shall see, it is precisely this "animalcentric" anthropomorphism that is not only acceptable but of great value in science.

Uninfluenced by Actual Behavior

Continuity between childhood and adult interests is by no means universal. One category of scientists has for professional reasons thrown up a barrier between themselves and the animals they study. Fouts must have thought of them when formulating his "first commandment," and indeed he had a supervisor who zapped his chimpanzees with a cattle prod. Rather than with such cruelty—which, fortunately, is rare—I am more concerned here with closeness versus distance.

Psychologists of the so-called behaviorist school are against the attribution of mental states to animals and hence have traditionally objected to any kind of anthropomorphism. This stance is somewhat puzzling, since this school, founded in the 1920s, initially strove for a unified theory in which human and animal behavior were subject to exactly the same principles. All behavior was explained by conditioning, that is, as stimulated or inhibited by positive or negative outcomes. The behaviorists' goal of applying a single explanatory scheme to all organisms was laudable, and its rigorous experimental procedures remain useful today.

There was a fundamental problem, however: people untrained in the behaviorist doctrine were prepared to buy its premises in relation to animals but most definitely not in relation to themselves. Early behaviorists considered emotions mere illusions, and mental states as residing inaccessibly in a "black box." It was therefore considered softheaded to even mention these phenomena. There was an appealing logic and consistency to dealing purely with behavior, and nothing else, except that no sane person was willing to accept the irrelevancy of feelings and thoughts for the production of human behavior. This forced a strategic withdrawal: behaviorists maintained their radical position with regard to animals only. Although never admitting it in so many words, in effect they abandoned their unified theory and increasingly began to treat animals and people as different. Whereas the human species was granted a mental life (although one not nearly as influential as most of us think it is), animals were kept at the level of stimulus-response machines.[21]

It is perhaps no accident that behaviorism is an American school, whereas ethology, with its emphasis on biology and instinct, is European. The first school has an optimistic, transformational streak (we can learn to become anything we want to be), whereas the second school assumes a measure of predestination.[22] For ethologists, every species arrives on this earth with a number of inborn behavior patterns that undergo little modification by the environment. Thus, the spider does not need to learn how to construct a web. She is born with a battery of spinnerets (spinning tubes connected to silk glands) as well as a behavioral program that "instructs" her how to weave threads together.

Because of their simplicity, both views of behavior had tremendous appeal, and obviously both touched on the truth.

At the time, however, this was not fully recognized because many scholars felt that nature and nurture were to be chosen between rather than combined. Yet, even if there is much less polarization now, one fundamental difference remains. Ethologists and naturalists are interested in animals for their own sake, whereas behaviorists mostly focus on a few domesticated animals, such as rats and pigeons, as "models" of the species that we belong to. Thus, one school sees animals as an end in themselves, and the other rather as a means to an end, positioned somewhere along a linear progression from "lower" to "higher" forms.

Even if many modern behaviorists have adopted a more cognitive and evolutionary perspective, I have run into the attitudes just described often enough to have developed a serious allergy to them. One example from my own experience is the reaction to the first hints of peacemaking among primates.

In the mid 1970s, I discovered that chimpanzees kiss and embrace after a fight, and dubbed these encounters *reconciliations*. When a new student joined my team, I proposed that she collect information on this phenomenon. There was a small problem, however. Whereas I was affiliated with Utrecht University, the student came from Amsterdam, and all of her professors were psychologists in the behaviorist tradition.

My supervisor, Jan van Hooff, and I went to Amsterdam to meet them. There we discovered that the entire committee was convinced that such a thing as reconciliation could never exist in animals. They knew only rodents, and in my inno-

Reconciliations are common and conspicuous in chimpanzees. In the top frame, a young female, on the right, is getting too close to the newborn of a high-ranking female. After the mother has slapped her away, the young female is screaming at a distance, hitting herself in frustration. In the bottom frame, she has returned to the mother and receives a kiss on her nose. After this, she is tolerated at close range again. Conflict resolution has been demonstrated in a host of primates. (Drawing by the author, first published in 1979).

cence I was surprised that they even had an opinion about primates. That they didn't take me seriously was one thing—I was young—but they also ignored Jan, an international expert on chimpanzees. We figured that perhaps we could change their minds by inviting them to the Arnhem Zoo, where our chimpanzees lived: seeing apes up close might be an eye-opener. To this proposal, however, they replied in a way that baffles me to this day: "What good would it do to see the animals? It will be much easier to stay objective if we are not influenced by that."

True, the discovery at stake was not nearly as momentous as that of the earth revolving around the sun, but the reaction nevertheless reminded me of that of the seventeenth-century church fathers who refused to look through Galileo's telescope. Who knows what they would have been forced to admit! It is said that the ancient King of Sardis complained that "men's ears are less credulous than their eyes." Only here it was reversed: these people feared that their eyes might tell them something they didn't want to hear.

And so, what Shepard saw as the "deeper sense of cohesion that sustains our sanity" has been lost in certain areas of science. Apparently, the human mind—and the behaviorist tradition with it—is so vulnerable to anthropomorphism that seeing, hearing, and smelling actual animals is to be avoided except when we know what to expect. Needless to say, the favoring of a theoretical doctrine over firsthand encounters with the organism does not develop naturally out of childhood curiosity: it represents a rupture, a throwing away of what Wilson called his talisman.

The "Which Is Which?" Approach

Like every biologist, I learned that one needs to build up an extensive background knowledge before one can even begin to address detailed questions. As Lorenz put it, one needs to grasp the whole before one tries to grasp its parts:

> One cannot master set research tasks if one makes a single part the focus of interest. One must, rather, continuously dart from one part to another—in a way that appears extremely flighty and unscientific to some thinkers who place value on strictly logical sequences—and one's knowledge of each of the parts must advance at the same pace.[23]

In the study of animal behavior, this means following each and every move of the species one is interested in, preferably under a wide range of circumstances. Behavior makes sense only in the larger context of the animal's natural history, social organization, general temperament, adaptations to its environment, and so on. One cannot expect predators to react the same as prey, solitary animals the same as social ones, vision-oriented animals the same as those relying on sonar, and so on.

I came across an amusing illustration in the scientific literature of the pitfalls for those who fail to pay attention to the

whole animal. In 1979 Bruce Moore and Susan Stuttard reported a replication of a 1946 study widely cited as demonstrating the ability of cats to work their way out of a puzzle box,
a container whose door was operated by moving a rod. The
earlier study, done by Edwin Guthrie and George Horton,
documented in great detail how cats rubbed against the inte-
rior of the box with stereotyped movements. In the process,
they moved the rod and escaped. Guthrie and Horton had
deemed it significant that all the cats in the experiment
showed the same rubbing pattern, which they believed they
had taught the animals through the use of food as rewards.
This proved the power of conditioning.

When Moore and Stuttard repeated the experiment, their
cats' behavior struck them as nothing special. The cats performed the usual head-rubbing movements that all felines,
from ocelots to jaguars, use in greeting and courting. Domestic
cats often redirect these movements to inanimate objects, such
as the legs of a kitchen table. Moore and Stuttard showed that
food rewards were absolutely irrelevant: the only meaningful
factor for the cats in the box was the visibility of people.
Without training, every cat that saw people while in the box
rubbed its head, flank, and tail against the rod and got out.
Cats who didn't see people just sat there. Instead of a learning
experiment, the 1946 study had been a greeting experiment![24]

The lesson is painfully obvious: before testing an animal, one
needs to know a bit about its typical behavior. Yet behaviorists of
the old school thought of animals as interchangeable. They rea-

Laboratory approaches to animal behavior sometimes overlook natural patterns of behavior, such as the affectionate head-rubbing of cats against the leg of a favorite human companion. (Gravure based on a drawing by T. W. Wood in Charles Darwin's The Expression of Emotions in Man and Animals, *1872).*

soned that if the laws of learning are universal, one animal is as good as another. As B. F. Skinner, one of the founders of the discipline, bluntly put it in *A Case History in Scientific Method:* "Pigeon, rat, monkey, which is which? It doesn't matter."[25]

Behaviorism started losing its grip, and was forced to adopt the premises of evolutionary biology, with the discovery that learning is not the same for all situations and species. In the best-known example, John Garcia reported that rats, which normally link actions with effects only if one immediately follows the other, are able to learn to avoid foods that make them sick even with a delay of hours between consumption and the negative sensation. Apparently, animals are specialized learners, being best at contingencies important for survival. Needless to say, if natural selection has shaped what is being learned, and what is not, the interchangeability of animal species is in trouble.[26]

In the old view, differences in intelligence were nothing else than differences in learning ability. Some animals have larger brains than others, which means that they learn faster, but reward and punishment still motivate all animals. But is a monkey brain really no more than an expanded rat brain, and is the human brain no more than a large monkey brain? Wouldn't it be surprising if evolutionary adaptation affects every anatomical feature one can think of—from limbs and teeth to stomach, eyes, and lungs—except for the brain?[27] If this were the case, the species with the largest brain would be superior in every respect.

This is clearly not so. Pigeons, for example, do better than humans at mentally rotating visual images, and some birds have an amazing memory for the location of hidden objects. Clark's nutcrackers store up to 33,000 seeds in caches distributed over many square kilometers and find most of the caches again months later.[28] As someone who occasionally forgets where he has parked an item as large and significant as his car, I am impressed by these peanut-brained birds.

Smoke and Mirrors

Biologists readily accept that the ability to recall locations makes perfect sense for an animal that relies on stored food, but to this date such specializations annoy behaviorists. And so, when Gordon Gallup, a psychologist at the State University of New York at Albany, demonstrated in a 1970 article a cognitive gap between apes and all the rest of the animal kingdom, including monkeys, his findings were sufficiently upsetting that two generations of behaviorists have broken their teeth on them.[29]

Gallup noticed that chimpanzees and monkeys respond differently to mirrors. Like most other animals, a monkey reacts to its reflection as if it were a friend or enemy, whereas an ape appears to realize that the image in the mirror is itself. Chimpanzees soon use the mirror to inspect parts of their bodies that are normally out of sight, such as the inside of their mouths or (in the case of females) their swollen pink behinds. Anyone who has ever seen an ape do this realizes that the animal is not simply opening its mouth or turning around accidentally: the ape's eyes closely monitor the movements of its body in the mirror.

To corroborate the observational evidence, Gallup designed an elegant experiment in which chimpanzees needed a mirror to detect a small change in body appearance. Known as the mark test, the experiment consists of painting a dot above the eyebrow of an anesthetized animal. Once the animal wakes up, it is shown a mirror. It cannot see the dot directly, but can detect its presence only in the mirror. In these experiments, the ape would stare at the dot in the mirror, then bring a finger to the real dot on its own face and inspect the finger afterward—a clear sign that the animals linked their reflections to themselves.

Apart from humans and apes, no other animals have convincingly passed this test, despite valiant efforts by many scientists.

Gallup spoke provocatively of self-awareness, and of the mental uniqueness of the hominoids, the family of animals made up of apes and·humans. This triggered one of the greatest travesties in behavioral science: an attempt to demonstrate the same ability in pigeons. Surely, if pigeons have self-awareness, these critics reasoned, the quality can't be so special. In 1981, B. F. Skinner and colleagues reported that, with enough trials, food rewards, and patience, they had managed to get pigeons to recognize themselves in a mirror. The birds pecked at dots projected onto their bodies, dots the birds could not see directly because they had bibs around their necks. A marvel of conditioning, no doubt, but the experiment did not convince many people that what these birds were doing after extensive human intervention was the same as what chimpanzees do spontaneously, without any help. Furthermore, attempts to replicate the results have remained suspiciously unsuccessful.[30]

Fifteen years later, another skeptical behaviorist tried a different approach. Cecilia Heyes, who was making a name for herself in Great Britain as a critic of the growing field of ape-intelligence studies, zoomed in on apes' responses to mirrors. Without the benefit of familiarity with primates, she came up with the creative suggestion that self-recognition might be a by-product of the anesthesia that is part of the mark test. Perhaps a chimpanzee recovering from anesthesia has a tendency to touch its own face in a random manner that produces occasional contacts with the dot. What other scientists had interpreted as self-inspection guided by a mirror might be a mere accident.

Heyes's idea was quickly disproved by an experiment in which Daniel Povinelli and his colleagues at the New Iberia Research Center carefully recorded which areas of the face chimpanzees touched in the mark test, and how soon after recovery from the anesthesia. They found that the touching is far from random: it is specifically targeted at the marked areas, and it peaks right after the ape's exposure to the mirror.[31] This is, of course, exactly what the experts had been claiming all along, but now it was official.

What makes critics such as Heyes unfathomable to me is their total absence of humility when faced with a group of animals they have never worked with. Behaviorists really do believe that they can generalize from rats and pigeons to all other species. But their "which is which?" approach to the diversity of life, and their talk of higher and lower forms, is essentially pre-Darwinian: it ignores the fact that every animal is a unique product of natural selection in both body and mind. Only those scientists who try to learn everything there is to know about a particular animal have any chance of unlocking its secrets. All others will keep tripping over the cat.

And so we return to scientists who erect no artificial barriers between themselves and other life forms, who respect animals sufficiently to realize that they can catch only a glimpse of what makes them tick, and who are not afraid to

To test the claim that chimpanzees randomly touch their face after recovery from anesthesia in front of a mirror, Povinelli and co-workers compared the number of contacts made with marked areas (black) with those directed at control areas (white). It was found that chimpanzees specifically target the marked areas, meaning that they are able to link their mirror image with their face. (Drawing by Donna Bierschwale, courtesy of the University of Louisiana-Lafayette, Cognitive Evolution Group).

identify with them, project emotions onto them, or trust their own intuitions about them rather than relying on pre-conceived notions.

I often see a parallel with so-called computer geeks. In the same way that some kids love animals, others spend all their time clicking away at computers, playing electronic games, browsing the World Wide Web, testing software, and so on. A few lucky people with this inclination are now highly visible billionaires, but they didn't start out with wealth in mind. They were just obsessed with the technology. Similarly, ethologists and naturalists are driven by a power beyond their control to work with animals, watching them for inordinate amounts of time. Their science follows naturally. The only difference, sadly, is that they never get rich.

The study of animal behavior is among the oldest of human endeavors. As hunter-gatherers, our ancestors needed intimate knowledge of flora and fauna, including the habits of their prey as well as the animals that prey on humans. The human-animal relationship must have been relatively egalitarian during this period.[32] Hunters exercise little control: they need to anticipate the moves of their prey and are impressed by the animals' cunning if they escape. A more practical kind of knowledge became necessary when our ancestors took up agriculture and began to domesticate animals for food and muscle power. Animals became dependent on us and subservient to our will. Instead of anticipating their moves, we began to dictate them.

Both perspectives are recognizable today in the study of animal behavior, and, to be successful, we need both the observer/hunter and the experimenter/farmer. But whereas the first can exist without the second, the second gets into all sorts of trouble without the first.

Are We in Anthropodenial?

The human hunter anticipates the moves of his prey by attributing intentions and taking an anthropomorphic stance when it comes to what animals feel, think, or want. Somehow, this stance is highly effective in getting to know and predict animals. The reason it is in disrepute in certain scientific circles has a lot to do with the theme of this book, which is how we see ourselves in relation to nature. It is not, I will argue, because anthropomorphism interferes with science, but because it acknowledges continuity between humans and animals. In the Western tradition, this attitude is okay for children, but not for grown-ups.

In one of my explorations of this issue, I ended up in Greece with a distinguished group of philosophers, biologists, and psychologists.[33] The ancient Greeks believed that the center of the universe was right where they lived. On a sun-drenched tour of the temple ruins in the foothills of Mount Parnassus, near Delphi, we saw the *omphalos* (navel) of the world—a large stone in the shape of a beehive—which I couldn't resist patting like a long-lost friend. What better location to ponder humanity's position in the cosmos? We debated concepts such as the anthropic principle, according to which the presence of human life on earth explains why the universe is uniform in all of its directions. Next to this idea, the Greek illusion of being at the navel of the world looks almost innocent. The theme of our meeting, the problem of anthropomorphism, related very much to the self-absorbed attitude that has spawned such theories.

Anthropomorphism and anthropocentrism are never far apart: the first is partly a "problem" due to the second. This is evident if one considers which descriptions of animal behavior tend to get dismissed. Complaints about anthropomorphism are common, for example, when we say that an animal acts intentionally, that is, that it deliberately strives toward a goal. Granted, intentionality is a tricky concept, but it is so equally for humans and animals. Its presence is as about as hard to prove as its absence; hence, caution in relation to animals would be entirely acceptable if human behavior were held to the same standard. But, of course, this is not the case: cries of anthropomorphism are heard mainly when a ray of light hits a species other than our own.

Let me illustrate the problem with an everyday example. When guests arrive at the Field Station of the Yerkes Primate Center, near Atlanta, where I work, they usually pay a visit to my chimpanzees. Often our favorite troublemaker, a female named Georgia, hurries to the spigot to collect a mouthful of water before they arrive. She then casually mingles with the rest of the colony behind the mesh fence of their compound, and not even the sharpest observer will notice anything unusual. If necessary, Georgia will wait minutes with closed lips until the visitors come near. Then there will be shrieks, laughs, jumps, and sometimes falls when she suddenly sprays them.

Georgia performs this trick predictably, and I have known quite a few other apes that were good at surprising people, naive and otherwise. Heini Hediger, the great Swiss zoo biologist, recounts how even when he was fully prepared to meet the challenge, paying attention to the ape's every move, he nevertheless got drenched by an experienced old chimpanzee. I once found myself in a similar situation with Georgia. She had taken a drink from the spigot and was sneaking up to me. I looked straight into her eyes and pointed my finger at her, warning, in Dutch, "I have seen you!" She immediately stepped back, let some of the water fall from her mouth, and swallowed the rest. I certainly do not wish to claim that she understands Dutch, but she must have sensed that I knew what she was up to, and that I was not going to be an easy target.

Georgia's actions are most easily described in terms of human qualities such as intentions, awareness, and a taste for mischief. Yet some scientists feel that such language is to be avoided. Animals don't make decisions or have intentions; they respond on the basis of reward and punishment. In their view, Georgia was not "up to" anything when she spouted water on her victims. Far from planning and executing a naughty plot, she merely fell for the irresistible reward of human surprise and annoyance. Thus, whereas any person acting like her would be scolded, arrested, or held accountable, some scientists would declare Georgia innocent.

Such knee-jerk rejections of anthropomorphism usually rest on lack of reflection on how we humans go about understanding the world. Inevitably, we ourselves are both the beginning and end of such understanding. Anthropomorphism—the term is derived from the Greek for "human form"—has enjoyed a negative reputation ever since Xenophanes objected to Homer's poetry in 570 B.C. because it treated the gods as if they were people. How could we be so arrogant as to think that they should look like us? If horses could draw pictures, Xenophanes joked, they would no doubt make their gods look like horses. Hence the original meaning of anthropomorphism is that of misattribution of human qualities to nonhumans, or at least overestimation of the similarities between humans and nonhumans. Since nobody wants to be accused of any kind of misattribution or overestimation, this makes it sound as if anthropomorphism is to be avoided under all circumstances.

Modern opposition to anthropomorphism can be traced to Lloyd Morgan, a British psychologist, who dampened enthusiasm for liberal interpretations of animal behavior by formulating, in 1894, the perhaps most quoted statement in all of psychology: "In no case may we interpret an action as the outcome of the exercise of a higher psychical faculty, if it can be interpreted as the outcome of the exercise of one which stands lower on the psychological scale."[34] Generations of psychologists have repeated Morgan's Canon, taking it to mean that the safest assumption about animals is that they are blind actors in a play that only we understand. Yet Morgan himself never meant it this way: he didn't believe that animals are necessarily simpleminded. Taken aback by the one-sided appeals to his canon, he later added a rider according to which there is really nothing wrong with complex interpretations if an animal species has provided independent signs of high intelligence. Morgan thus encouraged scientists to consider a wide array of hypotheses in the case of mentally more advanced animals.[35]

Unfortunately, the rider is not nearly as well known as the canon itself. In a recent assault on the "delusions" of anthropomorphism in the behavioral sciences, John Kennedy proudly holds up the behaviorist tradition as the permanent victor over naive anthropomorphism. He confidently claims in *The New Anthropomorphism:* "Once a live issue, a butt for behaviorists, it [anthropomorphism] now gets little more than an occasional word of consensual disapproval." In almost the same breath, however, the author informs us that "anthropomorphic thinking about animal behaviour is built into us. We could not abandon it even if we wished to. Besides, we do not wish to."[36]

This seems illogical. On the one hand, anthropomorphism is part and parcel of the way the human mind works. On the other hand, we have all but won the battle against it. But how did we overcome an irresistible mode of thinking? Did we really manage to do so, or is this a behaviorist delusion?

Is it even desirable to suppress thoughts that come naturally to us? Why is it that we, in Kennedy's own words, "do not wish to" abandon anthropomorphism? Isn't it partly because, even though anthropomorphism carries the risk that we overestimate animal mental complexity, we are not entirely comfortable with the opposite either, which is to deliberately create a gap between ourselves and other animals? Since we feel a clear connection, we cannot in good conscience sweep the similarities under the rug. In other words, if anthropomorphism carries a risk, its opposite carries a risk, too. To give it a name, I propose *anthropodenial* for the a priori rejection of shared characteristics between humans and animals when in fact they may exist.

Those who are in anthropodenial try to build a brick wall between themselves and other animals. They carry on the tradition of French philosopher René Descartes, who declared that while humans possessed souls, animals were mere machines. Inspired by the pervasive human-animal dualism of the Judeo-Christian tradition, this view has no parallel in other religions or cultures. It also raises the question why, if we descended from automatons, we aren't automatons ourselves. How did we get to be different? Each time we must ask such a question, another brick is pulled out of the dividing wall. To me, this wall is beginning to look like a slice of Swiss cheese. I work on a daily basis with animals from which it is about as hard to distance oneself as from Lucy, the Australopithecus fossil. All indications are that the main difference between Lucy and modern apes resided in her hips rather than her cranium. Surely we all owe Lucy the respect due an ancestor— and if so, does not this force a different look at the apes?

If Georgia the chimpanzee acts in a way that in any human would be considered deliberately deceitful, we need compelling evidence to the contrary before we say that, in fact, she was guided by different intentions, or worse, that apes have no intentions, and that Georgia was a mere water-spitting robot. Such a judgment would be possible only if behavior that in its finest details reminds us of our own—and that, moreover, is shown by an organism extremely close to us in anatomy and brain organization—somehow fundamentally differs from ours. It would mean that in the short evolutionary time that separates humans from chimpanzees, different motives and cognition have come to underlie similar behavior. What an awkward assumption, and how unparsimonious!

Isn't it far more economical to assume that if two closely related species act in a similar way, the underlying mental processes are similar, too? If wolves and coyotes have behavior patterns in common, the logical assumption is that these patterns mean the same thing, inasmuch as they derive from the common ancestor of both species. Applied to humans and their closest relatives, this rationale makes cognitive similarity the default position. In other words, given that the split between the ancestors of humans and chimpanzees is assumed to have occurred a mere five to six million years ago, anthropomorphism should be less of an issue than anthropodenial.[37]

This radical-sounding position—according to which, in the case of monkeys and apes, the burden of proof should be shifted from those who recognize similarity to those who deny it—is not exactly new. One of the strongest advocates of a unitary explanation was the philosopher David Hume. More than a century before both Lloyd Morgan and Darwin, Hume formulated the following touchstone in A Treatise of Human Nature:

> 'Tis from the resemblance of the external actions of animals to those we ourselves perform, that we judge their

internal likewise to resemble ours; and the same principle of reasoning, carry'd one step farther, will make us
conclude that since our internal actions resemble each
other, the causes, from which they are deriv'd, must also
be resembling. When any hypothesis, therefore, is ad-
vanc'd to explain a mental operation, which is common
to men and beasts, we must apply the same hypothesis to
both.[38]

Bambification

As soon as we admit that animals are not machines, that they
are more like us than like automations, then anthropodenial
becomes impossible and anthropomorphism inevitable. Nor
is anthropomorphism necessarily unscientific, unless it takes
one of the unscientific forms that popular culture bombards
us with. I was once struck by an advertisement for clean fuel
in which a grizzly bear had his arm around his mate's shoulders while both enjoyed a beautiful landscape. Since bears are
nearsighted and do not form pair-bonds, the image was nothing but our own behavior projected onto these animals.

Walt Disney made us forget that Mickey is a mouse and
Donald a duck. Sesame Street, the Muppet Show, Barney:
television is populated with talking and singing animal representations with little relation to their actual counterparts.
Popular depictions are often pedomorphic, that is, they follow
ethology's *Kindchenschema* (baby-appeal) by endowing ani-

mals with enlarged eyes and rounded infantile features designed to evoke endearment and protectiveness.

I've had firsthand experience with another form that I refer to
as satirical anthropomorphism, which exploits the reputation of
certain animals as stupid, stubborn, or funny in order to mock
people. When my book *Chimpanzee Politics* came out in
France in 1987, the publisher decided, unbeknownst to me, to
put François Mitterand and Jacques Chirac on the cover with a
grinning chimpanzee between them. I can only assume that he
wanted to imply that these politicians acted like "mere" apes.
Yet by doing so he went completely against the whole point of
my book, which was not to ridicule people but to show that
chimpanzees live in complex societies full of alliances and jockeying for power, societies that in some ways mirror our own.

You can hear similar attempts at anthropomorphic humor at
the monkey rock of most zoos. Isn't it interesting that an-
telopes, lions, and giraffes rarely elicit hilarity, but that people
who watch primates often end up hooting and yelling, scratching themselves in an exaggerated manner, and pointing at the
animals while shouting things like "I had to look twice, Larry,
I thought it was you"? In my mind, the laughter reflects anthropodenial: it is a nervous reaction caused by an uncomfortable resemblance.

The most common anthropomorphism, however, is the
naive kind that attributes human feelings and thoughts to animals based on insufficient information or wishful thinking. I
recall an interview with a woman in Wisconsin who claimed

that the squirrels in her backyard loved her to an extraordinary degree. The rodents visited her every day, came indoors, and accepted food directly from her hand. She spent over a thousand dollars per year on nuts. When the interviewer discreetly suggested that perhaps the abundant goodies explained the animals' fondness of her, the woman denied any connection.

Naive anthropomorphism makes us exclaim "He must be the daddy!" when an adult male animal gently plays with a youngster. We are the only animals, however, with the concept of paternity as a basis of fatherhood. Other animals can be fathers—and fathers may treat juveniles differently than non-fathers—but this is never based on an explicit understanding of the link between sex and reproduction. Similarly, when Elizabeth Marshall Thomas tells us in *The Hidden Life of Dogs* that virgin bitches "save" themselves for future "husbands," she assumes Victorian values in an animal not particularly known for its sexual fidelity.

All such instances of anthropomorphism are profoundly anthropocentric. The talking animals on television, the satirical depiction of public figures, and the naïve attribution of human qualities to animals have little to do with what we know about the animals themselves. In a tradition going back to the folktales, Aesop, and La Fontaine, this kind of anthropomorphism serves human purposes: to mock, educate, moralize, and entertain. Most of it further satisfies the picture, cherished by many, of the animal kingdom as a peaceable and cozy paradise. The fact that, in reality, animals kill and devour each

other, die of starvation and disease, or are indifferent to each
other, does not fit the idealized image. The entertainment industry's massive attempt to strip animals of their nasty side has
been aptly labeled their "Bambification."[39]

The general public is less and less aware of the discrepancy
with the real world as fewer people grow up on farms or otherwise close to nature. Even though having a pet provides a reality check (dogs are generally nice, but neither to their prey nor
to invaders of their territory), the full picture of nature in all its
glory and horror escapes the modern city dweller.

What Is It Like to Be a Bat?

The goal of the student of animal behavior is rarely a mere
projection of human experiences onto the animal. Instead,
the goal is to interpret behavior within the wider context of a
species' habits and natural history.

Without experience with primates, one might think that a
grinning rhesus monkey must be delighted, or that a chimpanzee running toward another with loud grunts must be in
an aggressive mood. But primatologists know from hours of
watching that rhesus bare their teeth when intimidated and
that chimpanzees often grunt when they meet and embrace.
In other words, a grinning rhesus monkey signals submission,
and grunting by a chimpanzee serves as a greeting. In this way
the careful observer arrives at an informed anthropomorphism
that is often at odds with extrapolations from human behavior.

When Sofie, a six-month-old kitten, bounces toward me sideways, with wide eyes, arched back, and fluffy tail, I recognize this as playful bluff. This judgment is not based on knowing any people who act this way. I just know how Sofie's behavior fits with all the other things cats do. By the same token, when an animal keeper says "Yummy!" while feeding mealworms to a squirrel monkey, she is speaking for the animal, not for herself.

Or take an example that reached the best-sellers list. In *The Man Who Listens to Horses*, animal trainer Monty Roberts freely employs what appears to be hopelessly anthropomorphic language to describe his animals' reactions. When the horses make licking and chewing movements, for example, they are said to be negotiating with their trainer: "I am a herbivore; I am a grazer, and I'm making this eating action with my mouth now because I'm considering whether or not to trust you. Help me out with that decision, can you, please?"[40]

Rather than attributing human tendencies to his animals, however, Roberts's interpretations are from the animal's perspective. His extraordinary success as a trainer rests on the fact that he treats the horse as a flight animal in need of trusting relations. A horse has a fear-based psychology totally different from that of a predator.

While the goal of understanding animals from the inside out may be considered naïve, it certainly is not anthropocentric. Ideally, we understand animals based on what we know about their *Umwelt*—a German term introduced in 1909 by Jacob von Uexküll for the environment as perceived by the animal. In the same way that parents learn to see through their children's eyes, the empathic observer learns what is important to his or her animals, what frightens them, under which circumstances they feel at ease, and so on.

Is it really anthropomorphic to look at the world from the animal's viewpoint, taking its Umwelt, intelligence, and natural tendencies into account? If anthropomorphism is defined as the attribution of human mental experiences to animals, then, strictly speaking, Roberts is not anthropomorphizing; he explicitly postulates major differences in the psychological makeup of horses and people. Although he does put human words in the horse's mouth, this seems done for the sake of reaching an audience, not because of any confusion between the species.

The animalcentric approach is not easy to apply to every animal: some are more like us than others. The problem of sharing the experiences of organisms that rely on different senses was expressed most famously by the philosopher Thomas Nagel when he asked, "What is it like to be a bat?"[41] A bat perceives its world in pulses of reflected sound, something that we creatures of vision have a hard time imagining. Still, Nagel's answer to his own question—that we will never know—may have been overly pessimistic. Some blind persons manage to avoid collisions with objects by means of a crude form of echolocation.[42]

Perhaps even more alien would be the experience of an animal such as the star-nosed mole. With its twenty-two pink, writhing tentacles around its nostrils, it is able to feel microscopic textures on small objects in the mud with the keenest sense of touch of any animal on earth. Humans can barely imagine this creature's Umwelt. Obviously, the closer a species is to us, the easier it is to do so. This is why anthropomorphism is not only tempting in the case of apes, but also hard to reject on the grounds that we cannot know how they perceive the world. Their sensory systems are essentially the same as ours.

Animalcentric anthropomorphism must be sharply distinguished from anthropocentric anthropomorphism (see diagram). The first takes the animal's perspective, the second takes ours. It is a bit like people we all know, who buy us presents that they think *we* like versus people who buy us presents that *they* like. The latter have not yet reached a mature form of empathy, and perhaps never will.[13]

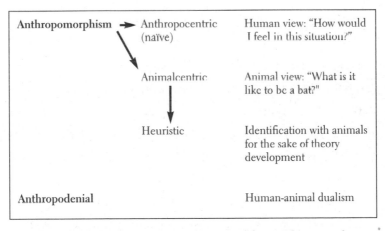

Anthropomorphism	→	Anthropocentric (naïve)	Human view: "How would I feel in this situation?"
		Animalcentric	Animal view: "What is it like to be a bat?"
		Heuristic	Identification with animals for the sake of theory development
Anthropodenial			Human-animal dualism

Anthropomorphism comes in many shapes and forms. The type to be treated with caution is the naïve, humanizing (anthropocentric) type. Most students of animal behavior, however, try to understand animals on their own terms. Animalcentric anthropomorphism is a common heuristic tool: it generates testable ideas. The opposite of anthropomorphism is anthropodenial, which is based on the assumption that it is safer to err on the side of difference than continuity.

To make proper use of anthropomorphism we must view it as a means rather than an end. It should not be our goal to find some quality in an animal that is precisely equivalent to some aspect of our own inner lives. Rather, we should use the fact that we are animals to develop ideas we can put to a test. This heuristic use of anthropomorphism is very similar to the role of intuition in all of science. It inspires us to make predictions, and to ask ourselves how they can be tested, how we can demonstrate what we think is going on. In this way, a speculation is turned into a challenge.[44]

Gorilla Saves Boy

On August 16, 1996, an ape saved a three-year-old boy. The child, who had fallen six meters into the primate exhibit at Chicago's Brookfield Zoo, was scooped up and carried to safety by Binti Jua, an eight-year-old western lowland gorilla. The gorilla sat down on a log in a stream, cradling the boy in her lap, giving him a gentle back-pat before continuing on her way. Her act of sympathy touched many hearts, making Binti a celebrity overnight. It must have been the first time in U.S. history that a gorilla figured in the speeches of leading politicians, who held her up as an example of much-needed compassion. *Time* elected her one of the "best people" of 1996.

Some scientists were not as lyrical. They cautioned that Binti's motives might have been less noble than they seemed, pointing out that she had been raised by people, and had been taught parental skills with a stuffed animal. The whole affair, they suggested, might be explained by a confused maternal instinct. Other speculations included that Binti might have acted the same way with a sack of flour, or that she presented the child to the keepers with the same "pride" with which a house cat presents a dead mouse to her owner.

The intriguing thing about this flurry of creative explanations was that nobody thinks of raising similar doubts when a person saves a dog hit by a car. The rescuer might have grown up around a kennel, have been praised for being kind to animals, and have a nurturing personality, yet we would still see his behavior as an act of caring. Why, then, was Binti's background held against her?[15]

I am not saying that I can look into Binti's heart, but I do know that no one had prepared her for this specific, unique emergency, and that it is highly unlikely that she, with her own seventeen-month-old infant riding on her back, was "maternally confused." How in the world could such a highly intelligent animal mistake a blond boy in sneakers and a red T-shirt for a juvenile of her species? Actually, the biggest surprise was how surprised most people were. Students of ape behavior did not feel that Binti had done anything unusual. Jürg Hess, a Swiss gorilla expert, put it most bluntly in an interview in *Stern*: "The incident can be sensational only for people who don't know a thing about gorillas."

What Hess meant—and I fully agree—is that Binti's action made a deep impression only because it benefited a member of our own species. To take care of a hurt juvenile is perfectly normal behavior for an ape, but of course it typically is di-

IF A POOR CHILD FELL THROUGH THE WELFARE SAFETY NET
WHICH ONE WOULD YOU TRUST TO RUSH TO HIS AID?

8·20·96 THE CHATTANOOGA TIMES

*Binti's 1996 rescue of a boy who had fallen into her enclosure at the
Brookfield Zoo, in Chicago, occurred when the United States was
gearing up for national elections with Bill Clinton, Bob Dole, and
Ross Perot as presidential candidates. (Cartoon by Bruce Plante for
the Chattanooga Times, with the artist's permission).*

rected at the ape's own kind. Instances of such caretaking behavior never reach the media, but they are well known and in
line with Binti's assistance to the unfortunate boy. The idea
that apes have a capacity for empathy is further supported by
how they embrace and caress recent victims of aggression, a
reaction thus far not observed in other primates.[46]

The incident at the Brookfield Zoo illustrates how hard it is
to avoid both anthropodenial and anthropomorphism at the
same time: in shying away from anthropomorphism one runs
straight into the problem that Binti's actions hardly make any

sense if one refuses to assume intentions and emotions. All one can come up with then is a confused instinct.

The larger question behind all of this is what kind of risk we are willing to take: the risk of underestimating animal mental life or the risk of overestimating it? There is no simple answer, but from an evolutionary perspective, Binti's kindness, like Georgia's mischievousness, is most parsimoniously explained in the same way that we explain our own behavior.

Darwistotle

The debate about anthropomorphism exposes old fault lines in Occidental thought, going back to the united view of Aristotle and the dualistic position of the Christian religion. Aristotle saw human social and political life as flowing from natural impulses, such as the reliance on cooperation and need for parental care that we share with many other animals. These views agree so well with current evolutionary biology that in the writings of at least one American political scientist, Larry Arnhart, Darwin and Aristotle have begun to blend into a single person, perhaps to be called Darwistotle.[47]

The Catholic Church, on the other hand, saw the universe as vertically arranged between heaven and earth. From this perspective, it made sense to speak of "higher" and "lower" forms of life, with humans being closest to the deity. Via philosophy, this way of thinking permeated all of the social sciences and humanities, where it still lingers even though biology has made it absolutely clear that the idea of a linear progression among life forms is mistaken. Every organism fits on the phylogenetic tree without being above or below anything else. Biologists make distinctions between organisms that do well or are extinct, that are specialized or generalized, or that multiply slowly or rapidly, but they never look at one organism as a model that others strive for or that is inherently superior.

These distinct strands of thought have made the West's relation with nature fundamentally schizophrenic: the dominant religion tells us that we are separate, yet science puts us squarely inside nature. This bears on the issue of anthropomorphism and explains why I attribute opposition to it to a desire to keep animals at arm's length rather than concerns about scientific objectivity. The latter is largely a rationalization by behaviorists.

Of course, picking on behaviorists is nothing new for an ethologist such as myself. Given the historical background of the contrasting views of humans' relationships to animals, there is nothing more logical than that an approach to animal behavior coming out of biology should collide with one coming out of psychology.[48] Behaviorists and ethologists have been at each other's throats ever since Daniel Lehrman wrote his stinging attack on Lorenz's instinct theory, in 1953. The ill feelings were mutual, and the stage seemed set for a disastrous split. That this didn't happen is due to both camps discovering in time that they shared a fascination with animals. Gerard Baerends, a Dutch ethologist, described his meeting with the enemy in Montreal, in 1954:

Jan van Iersel and I were the first ethologists to meet Danny Lehrman after his critique had appeared. After a few minutes of irrelevant behavior, we happened to discover that bird-watching was our common interest. This greatly facilitated our further exchange of ideas, which—as we soon found out—were far more compatible than earlier thought.[49]

Since then, both ethology and behaviorism have been transformed from within and now go largely by different names. The most effective criticism of behaviorist positions was delivered in-house by psychologists who questioned the ladder-like *Scala Naturae* view of evolution and objected to reliance on the albino rat as the animal of choice. Behaviorism still exists, but the old type has been relegated to history as "radical."[50]

The descendants of behaviorists call themselves comparative psychologists, a school that has grown considerably closer to ethology since Robert Hinde's grand synthesis, *Animal Behavior*, was published in 1966. The *Journal of Comparative Psychology* has become a meeting ground of students of animal behavior from backgrounds as diverse as traditional learning psychology, cognitive psychology, anthropology, behavioral ecology, and ethology. Thus, a recent handbook entitled *Comparative Psychology* lists Darwin, Lorenz, and Tinbergen among the discipline's pioneers even though all three were biologists.[51] The evolutionary approach and its attention to species diversity are clearly gaining ground, whereas opposition to mentalistic interpretations of animal behavior is more and more a rear-guard movement.

In the meantime, ethology, with its methodology of careful description and observation, has been absorbed into fields as diverse as child behavior and sociobiology. Even though early sociobiologists were quick to distance themselves from ethology as a way of showing that they were on to some new ideas (which was indeed the case), they were heavily inspired by it, especially by the Tinbergian school and its work on behavioral adaptation. Later, the term "sociobiology" fell into disrepute, leading to further name changes such as "behavioral ecology" and "evolutionary psychology," but without corresponding changes in outlook or research agenda.[52]

Whatever names we choose, students of animal behavior are still being trained largely in either biology or psychology, and the two have learned from each other, and have grown closer as a result. With increasing awareness of the flexibility of animal behavior, "instinct" is a term hardly used anymore, and the current interest in animal culture could be regarded as a triumph of those who have insisted all along on the importance of learning. In order to develop the study of behavior into a mature science we now need to find our inspiration in the Aristotelian view and organize our study along topic areas (such as cognition, evolutionary adaptation, culture, and genetics) rather than basing the structure of our discipline on whether we're dealing with a single bipedal primate or any other animal. Removal of this artificial split will go a long way toward calming the excessive fear of anthropomorphism, which fear was born from it.

2

The Fate of Gurus

When Silverbacks Become Stumbling Blocks

"Science is not about being right in the end. Much more important is to conquer unimagined fields, to stimulate, dare new approaches, engender debate. And this, Konrad Lorenz has done like no one else in this branch of biology."

Erik Zimen, 1999[53]

"Japanese culture does not emphasize the difference between people and animals and so is relatively free from the spell of anti-anthropomorphism . . . we feel that this has led to many important discoveries."

Jun'ichiro Itani, 1985

*I*nstead of marching onward with perfect vision, science stumbles along behind leaders who occasionally take the wrong alley, after which it turns to other leaders who seem to know the way, then corrects itself again, until sufficient progress is made for the next generation to either thrust aside or build upon. In hindsight, the path taken may look straight, running from ignorance to profound insight, but only because our memory for dead ends is so much worse than that of a rat in a maze.

Not surprisingly, leaders are treated with ambivalence. With the exception of those who have come up with absolutely invaluable insights, such as Einstein and Darwin, leaders first inspire and stimulate, then guide and protect their followers, but usually end up stifling further progress. They become major obstacles: the dinosaurs of fields that they themselves helped create. Hence the ugly patricides in which a number of upstarts revolt and get rid of the old guru.[54] They never do so literally, of course, but instead wield the academic version of the long knife, such as disparaging jokes during lectures, critical footnotes, bad book reviews, and after all is said and done, deadly silence.

Even though I will discuss two specific examples from the past—both related to animal behavior research in different cultural settings—the process should not be thought of as unique to any period. One patricide is going on as I write, perpetrated by a new generation of Darwinists who feel that the most popular evolutionary writer in the United States, Stephen Jay Gould, has been around long enough and holds old-fashioned, even erroneous views. Nasty lectures have been given, mean letters have been published, and in the ultimate

insult, Gould—who as no other has stood up against creation-
ism—has been labeled an "accidental creationist." The guru
himself is putting up an impressive fight, countercharging that
one of his opponents is the lapdog of another prominent evolutionist, facetiously apologizing to another for never having
heard of him, and lumping all of his adversaries together as
"fundamentalists." It isn't a pretty sight and reminds one of the
love affair with the guillotine after the French Revolution.
Now that the Darwinian approach is winning, at least in
academia, evolutionary revolutionaries feel the need for a
cleanup within their own ranks. And who better to drag to the
scaffold than the evolutionist most widely recognized outside
their little circle?[55]

Scientific leaders are often national figures, injecting entire
fields with cultural themes and reflections that may not be understood by outsiders. For example, Gould's fight against cre-
ationism is largely of local, North American significance, as
are his abundant references to baseball. Similarly, the two gurus to be discussed here were products of their respective cultures, which both happen to be defeated military powers. At
the peak of his fame, Konrad Zacharias Lorenz (1903–1989)
was the world's most admired knower of animals, the Dr.
Dolittle who told us how deep the similarities ran and warned
against the then popular *tabula rasa* view of human nature.
He had plenty of adversaries outside of biology—which never
bothered him—but later on his ideas increasingly met with resistance from within. Combined with reports about his

wartime past, this intramural antipathy made him vanish from the scene faster and more completely than any of his followers could have imagined.

Kinji Imanishi (1902–1992) is much less well-known in the West, but he was a towering figure in postwar Japan, where he instilled great national pride as a thinker and scientist, and inspired a whole generation to pursue the sort of primate studies that are at the heart of this book. Inasmuch as a sharp dividing line between humans and animals is not part of Oriental philosophy, and anthropomorphism was never seen as a problem, there were fewer impediments in place than in the West to conceive culture as applicable to other animals. But Imanishi, too, went under the ax of history and is now out of favor with some, and forgotten by most others, even in his own country.

By exploring the rise and fall of these figures we peek into the kitchen of science, and the cultural context within which it is conducted. The idea that science occurs in a vacuum never applies, not even to studies of animal behavior—or perhaps I should say, especially not to animal studies. We look at animals as informing us about ourselves, an orientation that sets the stage for the projection of values and the drawing of moral lessons. Scientists with the authority to do so play a different role, and enjoy a different status in society, than, say, prominent physicists or mathematicians. They are the high priests of nature who tell us where we came from and how we fit into the larger scheme of life.

All the more reason to pay attention to how these leaders affect history, and how history judges them. The latter is not to be left to their immediate successors, who often have political axes to grind. Even if ninety percent of their ideas are attributable to the old guard, they will praise to heaven the ten percent that they came up with themselves, sharply criticizing their predecessors for not having thought of it. Subsequent generations have the task of rediscovering the earlier contributions, which in the case of Lorenz and Imanishi are indisputably immense.

Coming to Terms with Konrad Lorenz

"Telling the truth isn't enough. The whole truth needs to be told. Nothing kept secret. This is where the responsibility of the scientist is greatest. He must not let anything of what he suspects about the possible applications or threats fall in the shadows."

François Jacob, 1998

An older social psychologist once shocked me by reacting to my declaration that I was a European ethologist with "So, you must be a Nazi!" I'd never been called a Nazi before, nor have I since: it is not an insult often leveled at Dutchmen. After all, my country submitted to Hitler only after he had laid waste to Rotterdam, and the Dutch never accepted his view that— since many of us are tall, blond, and blue-eyed—we belonged to the chosen race and should enthusiastically embrace the German cause in the same way as the Austrians had done two years before. Coming from a short, swarthy guy with a black moustache, the adoration of the Nordic phenotype had a certain comical quality. The Dutch, however, were not amused, and many resisted the occupation tooth and nail. One who failed to submit was Niko Tinbergen, the foremost Dutch ethologist, but the remark aimed at me referred not to him, but to Lorenz, the other founder of ethology.

I was offended, and went to the library to look up the founding fathers of the fine discipline to which this unfriendly psychologist belonged. What I found were books such as *Walden Two* and *Beyond Freedom and Dignity* by B. F. Skinner, and the writings of John B. Watson. The denial of thoughts and feelings (instead of loving each other, people were said to exhibit "conditioned love responses"), combined with the goal of complete behavioral control, made for a perfectly Orwellian worldview. For example, Watson proposed to hot-wire all items that children are not supposed to touch around the house ("I should like to make the objects and situations of life build in their own negative reactions"),[56] and Skinner raised his own daughter during the first years of life in his infamous Air-Crib. Skinner advocated training programs so as to produce ideal people who serve the greater good. He wanted to get rid of antisocial tendencies and dictate people's desires. All of us, he thought, should want what the social engineers have decided is best for our society.

But, of course, this was not the right way to deal with the issue at hand: it meant only that scary, totalitarian ideology could be found elsewhere than in biology. The real challenge was to see what, if anything, was the matter with Lorenz. I had

Like Konrad Lorenz, I am a jackdaw aficionado, having raised them by hand. Here a female, Rafia, is shown at the age of respectively 14, 26, and 37 days. (Drawing by the author)

become used to oblique remarks in the literature about his past, usually a mixture of embarrassment and apology, as if we would all be better off if history were left alone. But what if this wasn't as easy as it seemed to people who no doubt had a vested interest in letting sleeping dogs lie?

As a student I had avidly read Lorenz's first scientific paper, published in 1933, about jackdaws. A medium-sized member of the crow family, the jackdaw is my favorite bird. I raised and tamed them as a child, and kept several as a student—not in a cage, but freely flying in and out of my apartment window. I fed them as hungry fledglings every fifteen minutes for weeks until they were strong enough to take wing, which I "taught" them by throwing them high up in the air and letting them spiral down to my shoulder. They began to follow me on strolls through the neighborhood. I ended up developing a personal bond with each one of them, in some cases as their chosen mate. I also studied wild jackdaws in my first research project at the University of Groningen, climbing ancient steeples in which jackdaw colonies often nest.

Compared to most birds, who sail or flap through the skies on their way from A to B, the jackdaw is a merry customer who seems to thoroughly enjoy the incredible gift of flight, which we can appreciate only vicariously. In Lorenz's romantic prose:

> In the chimney the autumn wind sings the song of the elements, and the old firs before my study window wave excitedly with their arms and sing so loudly in chorus that I can hear their singing melody through the double panes. Suddenly, from above, a dozen black, streamlined projectiles shoot across the piece of clouded sky for which my window forms a frame. Heavily as stones they fall, fall to the tops of the firs where they suddenly sprout wings, become birds and then light feather rags that the storm seizes and whirls out of my line of vision, more rapidly than they were borne into it. . . .
>
> At first sight, you, poor human being, think that the storm is playing with the birds, like a cat with a mouse, but soon you see, with astonishment, that it is the fury of the elements that here plays the role of the mouse and that the jackdaws are treating the storm exactly as the cat its unfortunate victim.[57]

Lorenz described the behavior of these smart, congenial birds in such detail, with so many unexpected insights, and so completely in line with my own experience, that he forever won my regard for his powers of observation. Here was a scientist I wanted to learn from! As I read his many other descriptions of animal behavior, it was as if a thousand pieces fell into place. I had watched animals all my life, sometimes the same animals as Lorenz, and his writings always enriched my knowledge and understanding.

I learned that the secret of observation is to ask the right questions, and that observation needs to be followed by speculation about causes, functions, and connections between events. The goal is to sharpen the observations to the point that one is not just watching animals for pleasure and general information, but because one wants specific answers to specific questions.

Lorenz's interest in natural behavior came at a time when American behaviorists were busy teaching laboratory animals all sorts of tricks in their so-called Skinner boxes. This work produced critical insights into learning processes, yet it ignored the readily observable fact that every animal is born with behavioral tendencies typical of its species. Many animals survive through behavior that is only secondarily affected by learning, such as the dam building of the beaver or the weaving of the weaver bird. Because such "instincts" can be as fixed as anatomical traits, it is possible to compare them across species to trace their phylogeny in the same way as we do with the beaks of finches or the hands of primates.

Lorenz did not work only on instincts, however. Indeed, one of the early triumphs of ethology was the discovery of a learning process known as imprinting. Hatchlings of certain birds, such as geese and chickens, do not automatically recognize their species. Normally, they learn to which species they belong by following their mother around, but in experiments they can be imprinted upon any moving object, such as a toy truck or a walking or swimming zoologist. Imprinting takes place spontaneously, without any of the usual rewards and punishments. The tendency to absorb specific information at a specific age became known as a learning predisposition: a preprogrammed seeking out of critical information.

Lorenz also single-handedly inspired a massive amount of research on aggressive behavior. In his best-known work, *On Aggression*, which appeared in 1963, he argued convincingly that humans are naturally violent, thus starting a public debate continued by Robert Ardrey, Desmond Morris, Edward Wilson, and many others. His claim provoked fierce opposition from anthropologists, psychologists, and social scientists, who went on to demonstrate the critical role of learning in human aggression. This was very important for our understanding, but it did of course nothing to counter Lorenz's argument: a role for nurture by no means excludes a role for nature. Not that Lorenz's views are now accepted. His idea that aggression is produced by an inner drive has been heavily criticized, because most of the time aggression seems to be triggered by outside circumstances. Lorenz also considered lethal violence against members of one's own species an abnormal human characteristic, whereas we now know that it is actually quite widespread in the animal kingdom. But even with these qualifications, Lorenz remains fundamentally right on the point that aggression is an innate human potential.

All of this contradicted the then prevailing notion that animals, including humans, are born blank slates. Instead, Mother Nature's handwriting seems all over them. Lorenz's insistence that we pay attention to the natural context of behavior, its innate character, its function, and its evolution was new and exciting, producing what has rightfully been called "a blast of fresh air and intellectual vigor into a blinkered laboratory setting."[58]

Behind the Mirror

A charismatic, colossal Austrian with a big beard and a thick accent when he spoke English, Konrad Lorenz dominated scientific conferences with his mesmeric lecture style, boyish humor, inimitable animal imitations, affability, and approachability. Many ethologists with reservations about his scientific contributions nevertheless remember him as larger than life, a charming, irresistible father figure. One can imagine how his followers worshipped him.[59]

He wrote in a manner that everyone could relate to. Being the chief engine behind the popularization of ethology, he related the behavior of animals to that of people in freely moving, engaging prose rich in anecdotes, metaphors, and philosophical asides. He called behavior "the liveliest aspect of all that lives," and made sure that no one was bored when he had their ear.

Philosophy was from the beginning part of his work. This is perhaps not so surprising for a man who, born in 1903, grew up in Vienna playing cowboy with Karl Popper, later to become one of this century's foremost philosophers of science. Before World War II, Lorenz occupied Immanuel Kant's chair of psychology at Königsberg, in Germany. Lorenz became interested in the Kantian concept of the a priori: the built-in way in which the human mind organizes reality. In his very first book, he explored the possibility that what Kant saw as two parallel universes—the world outside and the a priori blueprint in the mind—were actually connected. What if a priori schemata were a product of evolution, a set of instructions for the human mind to sift through the huge amounts of information it encounters?

This splendid idea fits well with current thinking, such as the neo-Chomskian concept of inborn templates for language acquisition. Linguistic information enters a prepared mind that sorts it into the right boxes. Children absorb this information at impressive speed, but—as with the imprinting of young birds—there exists an optimal age. After this age the ability is sharply reduced, and it becomes nearly impossible to attain native levels of language proficiency.

Lorenz was far ahead of his time. His ideas about the "back side of the mirror" (his metaphor for a priori knowledge) were already developed between 1944 and 1948, in a camp in Armenia. As a prisoner of war, he wrote everything down on hundreds of cut-up pieces of cement sack under the most trying circumstances, often with frozen fingers and an empty stomach. The text was lost for a long time, but fortunately surfaced again. Wrapped in a newspaper, it had been hidden in one of the far corners of his library. His daughter posthumously published "the Russian manuscript" in 1992.[60]

But how did Lorenz end up in a prison camp?

The Dark Side

Initially, Lorenz could not get employment due to the dominant influence of the Catholic Church in Austria, which strongly opposed evolutionary views. Without a salary, he worked on fish and birds in the same house, in Altenberg near Vienna, where he had been born to a prominent family. His father was a famous orthopedist who introduced procedures that are still in common use.

Already in the late 1930s, during the rise of Nazism, it became clear that Lorenz looked with admiration at the new situation in Germany, and that he made friends with scientists with similar "brown" sympathies, such as zoologist Otto Koehler. This connection probably helped him get a job in Germany. In 1940, Lorenz accepted a position in comparative psychology at Königsberg, which he owed Koehler. There were outcries of protest against his appointment, not for political reasons, but because he was a biologist entering a psychology department.

Until 1996, when Ute Deichmann published *Biologists under Hitler*, I had known these facts only vaguely. Now the evidence was succinctly reviewed by someone unwilling to pull any punches. It went well beyond what I had expected.

Like many people today, Lorenz believed he lived in a period of moral decline. He saw parallels with animal domestication, a comparison that became a staple of his writings on the subject. Liberated from the harsh selection forces of nature, domesticated animals tend to lose adaptive traits; they become unfit for survival on their own. Similarly, Lorenz believed, human civilization supports more and more "degenerative types," which multiply freely because of "their larger reproductive rates and their coarser competitive methods towards the fellow members of the species." He argued that in the same way that one cannot remove a cancer from the body without drastic measures, race-hygienic (*rassenhygienische*) defense mechanisms are required to eradicate asocial elements from society. Like the Nazis, he emphasized the population as a whole—the common people, or *das Volk*—which he compared with a body.

Let me illustrate the disturbing tone of his statements:[61]

> Just as healthy tissue generally treats tumor as identical . . .
> the healthy volkish body often does not "notice" how it is
> being pervaded by elements of decay.

Any attempt to reconstruct elements that have fallen out of their relationship with the whole is hopeless. Fortunately, their elimination is easier for the people's doctor [*Volksarzt*] and much less dangerous for the supra-individual organism than the operation of a surgeon is for the individual body. The great technical difficulty lies in recognizing them.

The racial idea as the foundation of our form of government has already accomplished a very great deal.

Our species-specific feeling for the beauty and ugliness of the fellow members of our species is most intimately related to the manifestations of decay that are caused by domestication and threaten our race. One can see in this feeling almost a distinction of species-preserving importance for the elimination of such manifestations of decay.

Between 1940 and 1943, Lorenz repeatedly called for a deliberate, scientific race policy in order to improve Volk and race through the elimination of defective and inferior elements. What Skinner later would propose to achieve through brainwashing, Lorenz felt required harsh selection procedures. Two years after the introduction of the yellow star, he urged people to "beware of those who are marked." Theodora Kalikow, who studied the relation between Lorenz's writing and Nazi ideology, concluded that there was mutual stimulation between the two, "a process of reciprocal legitimation, whereby Nazis lent political power to ideas which were already part of Lorenz's world view. . . which may help explain Lorenz's increasing emphasis on animal and human degeneration after 1938. Lorenz's 'scientific evangelism' may have moved him to try to explain and justify Nazi racial policies ethologically."[62]

This interpretation is perhaps far-fetched, as it remains unclear that Lorenz's ideas carried much weight in Nazi circles. There is also no evidence that Nazi ideology affected Lorenz's theories about animals in any way. But it is true that Lorenz volunteered dangerous opinions cloaked in scientific jargon. Moreover, he acted upon his sympathies. Almost as soon as it was possible to do so after Austria's Anschluss, he joined the National Socialist party and obtained a permit from the party's Office of Race Policy to give lectures. In 1941 he was drafted into military service. Stationed as a military psychologist in Posen, he became involved in research on the deleterious effects of interbreeding. The study concerned hundreds of offspring from mixed Polish-German marriages: did these hybrids (*Mischlinge*) lack in noble German traits, such as an aptitude for hard work? The ultimate goal was to evaluate people's psychological "worthiness," ranking them on a four-point scale. Those on the bottom were to be deported.

Lorenz was only marginally involved in this gruesome enterprise, however. While Deichmann concludes that his involvement casts doubt upon later claims that he had known nothing about Nazi intentions,[63] it must be noted that he had nothing to do with the planning of these projects, spent most of his time at another post in a military hospital, and was mentioned in publications only as an honorary (as opposed to central) member of the research team.[64]

In 1944 Lorenz was transferred as medical officer to the Soviet Union. In the same year, he fell into Russian captivity. In 1948, well after the war was over, he returned to Austria with two hand-raised birds—a lark and a starling—as well as his mammoth manuscript.

After the War

It is impossible to fairly judge a man's character during a tumultuous period in history that one hasn't experienced firsthand. Only the arrogant think that they would have been heroes, doing the right thing, all others realize the frailty of the human spirit under trying circumstances.

Lorenz's opinions, moreover, were typical of racist attitudes of psychologists, anthropologists, and biologists at the beginning of the twentieth century.[65] Talk of superior and inferior types was common, often accompanied by drawings purportedly showing how people considered inferior were close to monkeys. But whereas it is one thing to express opinions like these as a theoretical exercise, in the Germany of the late 1930s things were rapidly moving from theory to praxis. In this political climate, Lorenz did not *need* to write about the removal of parasitic elements from the population. He did not *need* to become a member of the party. There may have been pressures to conform, but there was no obligation for scientists to publicly legitimize Nazi race policies. We cannot blame Lorenz for believing in his theory of human degeneration, but one wonders how he could have been blind to its application, or potential application, in this charged atmosphere.[66]

Even if we buy the excuse that Lorenz didn't know until after the war what the Nazis had been up to, the fact remains that he made no effort to correct his earlier views once he knew the full story. Instead, he maintained until his death that he had been too busy with his research to pay much attention to what went on during the war. In an interview on the occasion of his eighty-fifth birthday, in 1988, he stated: "In fact, I always avoided all politics because I was absorbed with my concerns. I also shirked a confrontation with the Nazis in a very disgraceful way, I simply did not have the time for it. . . . I reproach myself for it."[67]

Niko Tinbergen

One can hardly imagine a starker contrast with Lorenz than his Dutch counterpart, Nikolaas Tinbergen, who was a few years younger. The personalities, scientific methods, and war experiences of the cofounders of ethology differed enormously.

Tinbergen conducted experiments to test evolutionary hypotheses. He was particularly interested in behavior as an adaptation to the environment. For example, he wondered why birds remove eggshells after their chicks have hatched.

They take off with an empty shell in their bill, and then drop it away from the nest. Do they make this effort because chicks hurt themselves at the sharp edges of shells? Or is it because eggs are camouflage-colored on the outside but not the inside (which is exposed in the shells)? The second function was confirmed: predators, such as crows, find eggs much more readily if an experimenter has placed empty shells next to them. The parent birds themselves do not need to learn this costly lesson: eggshell removal is an automatic response favored by natural selection because birds that do it have more surviving offspring than birds that do not.

With his interest in adaptation, elegant tests to compare alternative explanations, and emphasis on statistical evaluation, Tinbergen pioneered the modern scientific approach to animal behavior. He was a more systematic and careful scientist than Lorenz, who was more of an animal-loving visionary. Reading current issues of *Animal Behaviour*, the foremost journal in my field, it is easy to recognize Tinbergen's lasting legacy.

Walking between the two giants of ethology as a young student, Robert Hinde—himself later to become a highly respected ethologist—listened in on a conversation. At issue was how often one needs to have witnessed a particular behavior in a species before one can claim it to be typical. Lorenz ventured that five times would do, whereupon Tinbergen slapped him on the shoulder, laughing "But sure, Konrad, you have never hesitated to describe things that you've seen only once." Lorenz could not deny it.

Hinde found that this little exchange captured in a nutshell the difference between these two men.[68] The one had a descriptive approach, and was a wonderful teller of animal stories, whereas the other was an experimentalist who wanted to see many observations with all sorts of controls before he would claim to have found something. Lorenz himself generously characterized their relationship as follows: "If ever two research workers depended on each other and helped each other, it is the two of us. I am a good observer, but a miserable experimenter and Niko Tinbergen is ... the past master of putting very simple questions to nature, forcing her to give equally simple and unambiguous answers."[69]

Even if many now believe Tinbergen to have been the better scientist, for a young science to grow and develop one needs all kinds. Lorenz and Tinbergen not only complemented each other in many ways; they stimulated and needed each other, even though it must have been emotionally difficult for Tinbergen, who refused to speak German after the war, to pick up their relationship where they had left off. Tinbergen had spent two years in a hostage camp in the Netherlands because he was part of a group of professors at the University of Leiden who had resigned their positions in protest against German efforts to "cleanse" the faculty. Nevertheless, from postwar correspondence it is clear that Tinbergen made a conscious effort to bring back to life the collaboration and exchange that ethology needed in order to survive. It is doubtful that at the time he knew the full magnitude of Lorenz's Nazi sympathies, but he decided that "through the bonds of personal friendship I hope to persuade them [ethologists, including Germans] to agree to a new start towards international cooperation."[70]

Between Science and Ideology

Exoneration by influential scientists, such as Tinbergen, goes a long way toward explaining the rehabilitation of Lorenz. His past never became a major issue in ethological circles, even though it was often referred to in passing.[71] The most outspoken resistance concerned not Lorenz's support of race policies, but a theoretical point he made over and over in the course of his career: the concept of species preservation (*Arterhaltung*).

Lorenz believed that animals behave in certain ways for the sake of their species. For example, they restrain aggressive impulses because otherwise they might drive their species into extinction. This whole idea went out the window with the rise of sociobiology in the 1970s. Sociobiologists firmly opted for the individual as the unit of selection, seeing selection at the group level as a rather limited phenomenon. Individuals who care only about themselves will outcompete individuals who care for their group. This doesn't mean that the group is unimportant, but rather that it is important only insofar as it serves its members. Selection at the species level was not even seriously considered: there doesn't seem to be any reason why animals should care about their species.

The now outdated concept of species preservation is along the same lines as Lorenz's insistence, during the war years, that the individual is insignificant compared to the people (Lorenz literally said "Race and Volk are everything to us, the individual almost nothing"). By stressing the health and purity of *das Volk*, he placed the welfare of the whole above that of its parts. Such an outlook does have moral implications. A German anthropologist and primatologist, the late Christian Vogel, openly criticized this line of thought, warning against the value judgment implicit in the equation of "adaptive" with "good." If the definition of "good" refers to the condition of the population—as advocated by Lorenz—there is great risk that the moral rights of individuals, especially those who don't conform, will be trampled.

In this respect, sociobiology parted ways in no uncertain terms with Lorenzian ethology. Those who, in the 1970s, denounced sociobiology as a fascist theory got it all wrong. They were blinded by the possible link between biology and human behavior, and the dangers they felt this link represented. Sensitivity in this regard is fully understandable, yet fascist ideology is quite different. It stresses the supraindividual level, whereas sociobiology focuses on much smaller units, mostly individuals. This emphasis alone should render sociobiology unsuitable as a tool for totalitarian ideology.

Not that emphasis on the individual is free from political implications. It plays into the hands of conservatives, who love to argue that societies are artifacts, and that each individual should simply follow its own greedy genes.[72] Seeking to weaken the social contract, these ideologies make the opposite error from those that place the whole above its parts: they try to pry the parts loose from the whole.

Evolutionary biologists should face these debates head-on. We are so familiar with the tension between unity and diversity that we have no trouble exposing the flaws in both forest-for-the-trees and trees-for-the-forest ideologies. Surely, fascists have no evolutionary leg to stand on when they treat populations as organic wholes. Similarly, radical individualism is untenable in a world in which many species, including our own, survive through mutualism and cooperation. Both the pursuit of individual happiness and community-level morality have a place within evolutionary thought. We have an obligation to make sure that the metaphors and simplifications of our field are not used to downplay one or the other, and that it is understood where the realm of science ends and that of ideology begins.

Lorenz Today

One of Lorenz's former co-workers, Norbert Bischof, recently lamented that "It has become silent around Konrad Lorenz." It may be worse: Lorenz is increasingly disrespected. Even his compatriots have begun to minimize his contributions, as if all he did was tell amusing stories. Here is a condescending observation by Austrian philosopher and cultural commentator Konrad Liessmann:

What is admirable about him—of whom it is said that he would have preferred to be a greylag goose—is his empathy with the animal, his singular gift for observation, his talent to communicate with animals. . . . He may not survive as ground-breaking biologist, even less as cultural critic, but rather as author of enthralling animal tales. Perhaps they have given Konrad Lorenz the wrong Nobel Prize.[73]

Ever since reductionism gained the upper hand in my field, there has been great eagerness to dismiss Lorenz, and with him the holistic approach he called the *Ganzheitsbetrachtung*, or contemplation of the whole. Many consider him out of date, almost irrelevant. But those who present Lorenz's evolutionary ideas as the pinnacle of foolishness usually fail to add that he was by no means alone in adhering to a naive functionalism that emphasized phylogeny rather than natural selection. Yes, these views have been left behind, and rightly so, but if we were to dismiss every scientist who held views that are no longer accepted, the only remaining category would be us: those brilliant minds of the here and now.

The sharpest criticism of Lorenz has traditionally come from scientists who trace their background to Tinbergen. Yet, even though I greatly admire Tinbergen and consider myself his intellectual kin through the Dutch ethological family tree, the absolute reign of behaviorism would never have been broken by this gentle, thoughtful man on his own. He needed a warrior by his side, someone unafraid to call nonsense "nonsense." The case for animal behavior as a product of evolution was presented by Lorenz with such authority and conviction that no one could ever again look at a Skinner box without asking what lever pressing had to do with survival. While Tinbergen sought to convince audiences with well-designed experiments, Lorenz blew them away by the sheer force of his personality and the directness of his knowledge. Someone who swims with the geese and climbs onto roofs wearing a mask to fool the birds may indeed be wearing King Solomon's ring. No one who read or heard him doubted for a moment that Lorenz knew his animals inside out.

Edward Wilson was greatly affected. Even though he went to see both Tinbergen and Lorenz when they came to lecture at Harvard in 1953, it was Lorenz who made by far the deeper impression. Wilson describes him as "a prophet of the dais, passionate, angry, and importunate." Lorenz challenged prevailing views, saying that the role of learning was grossly overestimated, and that the answers to animal behavior were to be found in evolutionary biology. The young Wilson happily concluded that "Lorenz has returned animal behavior to natural history. My domain."[74]

This is not a contribution to be sneezed at. To throw an inextricable wrench into the wheels of behaviorism, to pave the way for the evolutionary study of behavior, and to conceive of the mind as a knowledge-acquisition device shaped by evolution are major accomplishments that reverberate to this day. Lorenz greatly influenced scientists of his time, pushing them to conduct ever more sophisticated research. A British contemporary, William Thorpe, rightly observed that "the very fact that Lorenz's early papers appear now so outdated is a tribute to their effectiveness."

How painful that a man of such obvious qualities, with such a profound understanding of animals, will forever have a cloud hanging over his head. His racist writings bother me far more than whatever brand of evolutionary ideas he held. Why is it that love for animals doesn't always translate into love for people? Lorenz saw civilization, like domestication, as a corruption of nature. And in the end, he helped fuel the fire that dealt civilization one of its biggest blows.

Since it is impossible to separate Lorenz the man from Lorenz the scientist, I may never be able to shake off my mixed feelings. On the other hand, there is absolutely no reason to doubt his contributions to the field of animal behavior: they have been profound and deserve our continuing respect and gratitude.

Kinji Imanishi and the Rabid Englishman

"In my Western way, I came to Kyoto, the home of Imanishi and his School seeking the man and his ideas, but I came as an avowed opponent."

Beverly Halstead, 1984[75]

In 1984, an eccentric Englishman, who couldn't resist comparing himself to a nineteenth-century explorer, landed on an Eastern shore. As if possessed, he hammered away day and night on an old typewriter until he had a rather disorganized product in his hands: a volume of over two hundred pages. Along with naïve comments on a society that he didn't seem to like, the rambling text defended Darwin against the dominant Japanese scientist of the day, Kinji Imanishi. All of this was accomplished in a one-month period, thus defying the old saw that in order to write about Japan one needs to stay either three weeks or thirty years.[76]

Beverly Halstead's colonial attitude was complete: a heavy load of prejudice about the country he was visiting, absence of knowledge about his adversary (all of Imanishi's important works are in Japanese, a language Halstead admitted not knowing), manipulation by the locals (the author had been invited to Japan by left-wing professors out to undermine Imanishi without getting their hands dirty), and earthshaking cultural discoveries, such as that the Japanese are more individualistic than one might think.

It is impossible for a Westerner to read Halstead's manuscript—dug up from a Kyoto library—without feeling one's toes curling in embarrassment, especially when one realizes that the text subsequently appeared in Japanese.[77] The Englishman didn't waste time on politeness. At one point, he managed to meet Imanishi in person, an opportunity he used to lecture the eighty-two-year-old emeritus professor. After having handed the father of Japanese primatology a gift—a bottle of whisky—he confronted him with a carefully translated document that included statements such as "Imanishi's evolution theory is Japanese in its unreality" and "You see the wood, but the trees are not in focus." No wonder Halstead describes Imanishi's facial expression on this occasion as one of profound regret at having agreed to the encounter.

What could possibly have compelled Halstead to be so rude? Why, upon return to his homeland, did he write an article that trashed not only Imanishi's views but an entire culture? How did *Nature* even dare to run it with a patronizing opening line like: "The popularity of Kinji Imanishi's writings in Japan gives an interesting insight into Japanese society"?[78] If the whole affair provides any insight at all, it is into Halstead's personality.

The late Beverly Halstead, from the University of Reading in Great Britain, was by training a geologist and paleontologist. Known for communist sympathies in his early years, he later became a flag bearer for Darwinism. Once described as "Darwin's Terrier" (in a play on T. H. Huxley as "Darwin's Bulldog"), Halstead had a professional life peppered with spectacular quarrels. An obituary in *The Independent* of May 3, 1991, highlights the nature of his combative attitude: "[He] was never the rebel but the supporter of traditional orthodoxy against what he saw as misplaced enthusiasm for the new." I guess he was the kind of person who sought security in doctrine—*any* doctrine. We all know the kind: the former Marxist who turns devout Catholic, or the people who escape the grasp of a sect only to become born-again Christians. Halstead was definitely not Christian ("Darwin rendered the entire edifice of Christianity redundant," he wrote), but he clearly thirsted for dogma.

To him, Imanishi's disagreement with Darwin was blasphemy. He came to set the old man straight, and with him an entire nation that, in his words, was engaged in a peculiar conspiracy to mislead everyone about themselves. The emphasis in Japan on social harmony is pure self-deception, Halstead concluded, because we all know that underneath there must exist incredible competition. Coming from a former communist, this was an interesting thought.

Imanishi as Founder

For Imanishi harmony was not an illusion, but at the core of all that lives. After having talked with dozens of Japanese colleagues about this man, not only in Kyoto, but all over the country, I began to joke that there were about one thousand different opinions out there. No two people thought the same about him. At one extreme, there was Jun'ichiro Itani, Imanishi's most influential student, whom I met over a delicious lunch at which the jumbo shrimps were so fresh that one literally jumped off the plate. The seventy-two-year-old Itani remains a great admirer of his late teacher. At the other extreme was the younger generation, especially Mariko Hiraiwa-Hasegawa, who believe Imanishi set back Japanese ecology and primatology for at least one, perhaps two decades by opposing the sociobiological revolution. Itani and Hiraiwa-

Hasegawa, each in their own way, have followed in Imanishi's footsteps by bringing biology to the masses. Both have written best-sellers, and both are well-known public figures in Japan.

Imanishi was an extraordinarily prolific, widely known author in the life sciences: the Stephen Jay Gould of Japan. He started out as an entomologist, but he was also an ecologist, anthropologist, primatologist, mountaineer, and philosopher. He received an official faculty position—in the humanities, not the sciences—only after he was about fifty. Coming from a wealthy family, he could do whatever he wanted without the obligations that come with salary. He had only one room at Kyoto University with no furniture other than a low desk at which he wrote his books sitting in lotus position on a *tatami*: an ascetic, cultured man of immense influence. Later in life, he would say that he was not a scientist, that his thinking was rather that of a poet.

Apart from being a pioneering Himalayas climber, Imanishi had two main interests. The first was the interconnectedness among all living things and the environments in which they are found. Even though he rarely mentioned those who influenced him, he was widely read, and elements of his approach are traceable to outside influences, ranging from Jacob von Uexküll to Petr Kropotkin, and perhaps most of all Kitaro Nishida, founder of a school of philosophy that was particularly influential in the 1930s and 40s. I cannot judge this for myself, as my information comes from secondary sources, but Imanishi's emphasis on intuition and perception of the whole, his dislike of reductionism, and his view that the individual is secondary to the society probably derived from Nishida, the Kyoto philosopher of "nothingness," who used to think deep thoughts while strolling along a rustic little river lined by cherry trees—still known as the philosopher's way—that runs past the university campus.[79]

Imanishi may have gotten wet in the same stream, although he did most of his research on mayfly larvae in the much larger and faster Kamogawa River, which runs through Kyoto's heart. His work on aquatic life led him to develop the idea of habitat segregation, meaning that different but related species select their own distinct lifestyles and micro habitats, which allows them to coexist harmoniously in the same environment. An associated idea was that organisms possess a species identity and form a species-level society, called *specia*, that controls individual behavior. When one species evolves into another, for example, the specia *en masse* opts for a different lifestyle. Hence the oft-repeated slogan of Imanishiism: "When the time comes, every individual will change simultaneously."

To my ears these are murky ideas. Species-level control over individuals? My first thought would be that it is the individual members of a species that optimize their life in a particular habitat, and that the success of a species follows from how well its members fare. As for habitat segregation, if two related organisms live peacefully side by side in different ecological niches, this doesn't necessarily mean that their initial parting of ways wasn't based on competition. Imanishi was vehemently opposed, however, to explanations that involved strife, and didn't seek to explain *how* segregation might have come about. Instead, he focused on the end result.[80]

Imanishi's second interest, and lasting legacy, is easier for me to judge, as it concerns the study of primate behavior. Here his approach was very innovative, thanks to the absence of human-animal dualism. Being the product of a culture that doesn't set the human species apart as the only one with a soul, Imanishi had trouble with neither the idea of evolution nor that of humans as descendants of apes. To the Buddhist and Confucian mind, both ideas are eminently plausible, even likely, and there is nothing insulting about them.[81] The smooth reception of this part of evolutionary theory—the continuity among all life forms—meant that questions about animal behavior were from the start uncontaminated by feelings of superiority and aversion to the attribution of emotions and intentions that paralyzed Western science. Thus, the Japanese did not hesitate to give each animal a name or to assume that each had a different identity and personality. Neither did they feel a need to avoid topics such as animal mental life and culture. Imanishi's students moved ahead rapidly with a distinctly anthropological agenda: by studying other primates, they sought to understand the origins of the human family and society. The presence of monkeys on their soil only helped, of course.

In all of this, Imanishi was well ahead of the celebrated paleontologist Louis Leakey, who developed a similar agenda. Leakey sent Jane Goodall and other primatologists out to study great apes in the wild in the belief that these animals could inform us about the earliest stages of human evolution. But by the time he did so, in the 1960s, the questions and techniques that would prove useful in this sort of endeavor had already been developed by Japanese primatologists, who had individually identified their monkeys and followed them long enough to understand the importance of kinship, the unexpected complexity of primate society, and the degree to which every group was different. Most importantly, as I discuss in Chapter 5, Imanishi had formulated the question of animal culture in a way that invited further study.

When Japanese primatologists went to Africa to observe great apes in their natural habitat, they arrived with excellent training and their hallmark approach of persistent, long-term data gathering that was to become the standard. Like Goodall, they habituated the objects of their study to human presence through food provisioning. Major discoveries were made by these scientists, such as that chimpanzees live in well-delineated groups, and that they use lithic tools that, had they been associated with people, would have qualified them for the Stone Age.[82]

But instead of comparing Imanishi with Leakey, the more appropriate parallel is with Ray Carpenter, the American primatological pioneer. Because Carpenter made no spectacular discoveries and wrote no popular books, he is now all but forgotten. Many, though, consider him the founder of Western primatology, together with experimental psychologists such as Wolfgang Köhler and Robert Yerkes. Carpenter was a trained physiologist, but also a first-rate behavioral scientist who preferred the field over the laboratory. He worked on rhesus macaques released on the Caribbean island of Cayo Santiago as well as on wild howler monkeys and gibbons. He was interested in social relationships and drew sociograms that mapped group structure. He didn't go nearly so far in this as the Japanese primatologists, who were able to tell over a hundred monkeys apart and trace their family ties over generations, but Carpenter shared with them a distinctly sociological outlook.

Carpenter identified individuals by means of tattoos, hence with an initial underestimation—typical for Western science of those days—of their individuality. It would be a bit like me going to a party and putting colored dots on everyone's foreheads saying that otherwise I couldn't tell these people apart. The party goers would be insulted, and rightly so. But in reading Carpenter, it is obvious that he was a most perceptive observer. Not surprisingly, when he first heard of the Japanese studies, he didn't share the skepticism of so many of his colleagues, who reacted with disbelief that monkeys could be distinguished by sight. They regarded all this naming of individuals as hopelessly anthropomorphic. Weren't the Japanese grossly overestimating the social lives of their monkeys, and who said that monkeys could tell each other apart even if human observers said *they* could?

Being perhaps the only one to fully appreciate the task Imanishi and his followers had set themselves, Carpenter became a big fan of their work. The older generation of Japanese scientists remember this gentle man with fondness and are

The Tibetan macaque is with over 20 kg the heaviest member of its genus, which also includes the better-known rhesus monkey and Japanese "snow" monkey. Tibetan macaques dwell in mountainous regions of China. Since Buddha is thought to have been a king monkey, the macaques are greatly respected. Here an adult male with an expressive bearded face. (Huang Shan, China, photo by author).

Male Tibetan macaques have such large testicles that sperm competition may have taken over from actual combat over mating rights (left). This could explain the relatively relaxed relations among males: they do occasionally fight and obey a strict hierarchy, but males of this species are also unusually close. The grin of the male above signals submission to a higher ranking male. Two males on the right scream with excitement while mounting each other: a common bonding ritual. (Huang Shan, China, photos by author).

Fifty years after the potato washing habit spread among Japanese monkeys on Koshima, they are still doing it. This is remarkable because for the last quarter century, they have received potatoes only a few times a year, and the current population has never known the innovator. How such habits spread is a point of debate. It is logical to expect that the infant clinging to its mother will learn to associate potatoes with the ocean simply by picking up dropped pieces. (Photo by author).

Top, view of the famous Koshima beach where cultural studies of animals found their origin. In the foreground, the freshwater stream in which Imo, a juvenile female, dipped her first sweet potato before other monkeys, beginning with her mother and peers, followed suit. It is only later that the monkeys moved their cleaning operation to the ocean, visible in the background. Bottom, two juveniles have carried potatoes to the salty ocean. (Photos by author).

I visited many primate field sites in Japan. Here I am watching a peaceful grooming scene at Katsuyama, where recently monkeys developed the habit of washing grass roots before eating them. (Photo by Catherine Marin). Below, a monkey at Arashiyama rubs two rocks together, a habit transmitted at this site for over twenty years. Cultural learning does not appear to follow the same rules as contingency learning in the laboratory: So far as we know, stone handling delivers no rewards. (Photo by author).

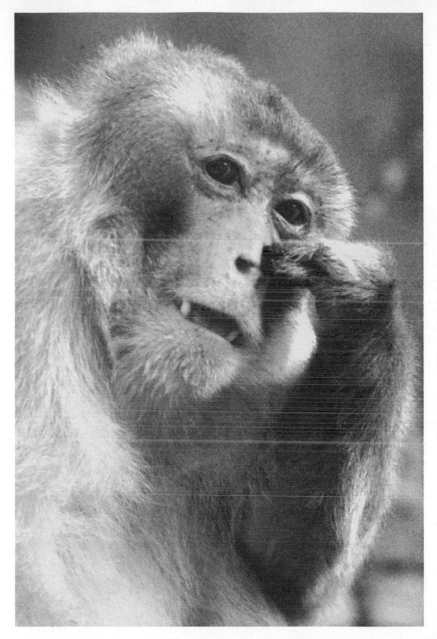

Mr. Spickles, who for many years seemed the absolute leader of a group of rhesus monkeys, was nevertheless dependent on female support, which kept him firmly in the saddle in the presence of stronger, younger males. (Wisconsin Primate Center, photo by the author).

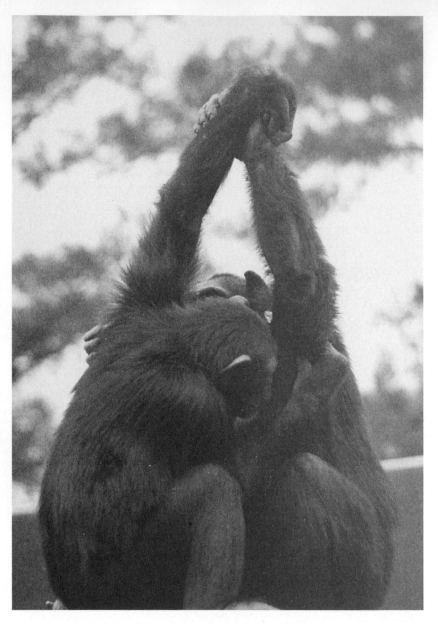

The only captive group of chimpanzees that knows the hand-clasp grooming posture, also observed in a few wild communities, resides at the Yerkes Primate Center's Field Station in Georgia. The custom has spread slowly from its originator, a female named Georgia, to virtually all adult members of the group: here between Georgia's sister and mother. (Photo by author).

puzzled why he never achieved fame in the West equivalent to Imanishi's fame in Japan.

Looking back at Imanishi's influence, it is clear why his ideas about harmony among species and his opposition to reductionism would create trouble with Darwinists. But we should not let these problems overshadow the enormous accomplishments of Imanishi's approach to primate behavior, which amount to a paradigmatic shift adopted by all of primatology and beyond. The basic premises of his school, and its application of ethnography to the study of animal societies, are now all but taken for granted. If students of long-lived animals in the field—whether they watch wolves or elephants—routinely identify individuals and follow them over their life spans, they are employing a technique from the East initially mocked and resisted by the West.[83]

In our own creative way, we have come to justify the need for this kind of observation, arguing that since natural selection operates on individual variation, every individual deserves special attention. But in order to make this step we had to overcome thousands of years of cultural prejudice against "lower" animals. We had to be shown the way by scientists immune to this hierarchical point of view.

Imanishi as Obstacle

To depict nature as harmonious, as Imanishi did, is entirely legitimate: there are indeed large segments characterized by peaceful coexistence, equilibrium, and symbiosis. To look for competition and tension is equally legitimate, and such an enterprise doesn't come up empty-handed either. What is silly, however, is to depict emphasis on harmony as culturally biased and emphasis on competition as somehow scientifically objective, as Halstead did. These are the remarks of a man unwilling or unable to examine his own perspective, which ironically shares with Japanese culture that it has been molded by life on a crowded island where people think they are unlike any other.

Imanishi's defenders have complained about Halstead's prejudices: "To understand how this seems to a Japanese, the Western reader might imagine the translated account of a Japanese professor, who speaks no English, who has spent a few months at Oxford and who gradually works out, through the characters he meets, Oxford's contradictions and Britain's hidden class structure."[84] In the same breath, the visitor might have noted—as Halstead did with Imanishi—how Darwin's theory reflects the society that produced it. That ideas about free-market capitalism and the struggle for life arose at the same time and in the same place can, of course, hardly be coincidental. Hence, when Halstead accuses Imanishi of being the product of a feudal society in denial of individualism, and when *Nature* prints this view along with a defense of Darwinism as the only way to understand the world, all we have is the familiar case of one culture perceiving another's biases more acutely than its own.

This is not to say that Imanishi didn't ask for a confrontation. Had he merely formulated a theory emphasizing the harmonious side of nature, no one would have protested, either in Japan or outside. But increasingly he began to present habitat segregation as a viable alternative to Darwinism, as incompatible with and surpassing traditional evolutionary views. As a result, he became positively hostile to Darwin. This stance invited direct comparisons between the power of Imanishi's theory and that of Darwin's. With great frankness, Hiraiwa-Hasegawa makes clear that this comparison was not in Imanishi's favor:

> He repeatedly explained that survival of the fittest doesn't explain anything, that such mechanical, purely biological explanations are not sufficient for a full understanding of organisms, that natural selection based on competition among individuals is a distorted view of the organic world from a Western point of view. He has written 30 volumes on his theory of evolution, and it is widely known as Imanishi's theory in Japan. However, none of his books provides mechanical or scientific explanations on how these things occur and how the theory can be tested. I think this is not science: this is a personal view on the organic world.[85]

Halstead drew a parallel with Kropotkin, who also selectively looked for and found examples of cooperation in nature. Yet throughout his life the Russian prince remained a staunch admirer and follower of Darwin. The target of Kropotkin's wrath was not Darwin, but Thomas Henry Huxley and his narrow outlook—often described as "gladiatorial"—that pitted every life form and every individual against every other. Kropotkin rightly noted that many animals survive not through struggle, but through mutual aid. Huxley's cardboard version of Darwinism was exactly the sort of caricature that Imanishi also objected to, and that he may have mistaken, quite erroneously, for Darwin's own perspective.

And so what remains is that Imanishi underestimated how his ideas, without too much trouble, could have been reconciled with Darwin's. Even though the process of natural selection is inherently competitive, it has produced all sorts of tendencies and configurations in nature, including socially positive and cooperative ones. The lethal territoriality of the male tiger is as much a product of natural selection as the death-defying solidarity among dolphins. Inasmuch as confusion between process and outcome has led scientists in the West to doubt that humans and animals can be genuinely nice, we shouldn't be too hard on Imanishi for making the opposite error, which was that he doubted the competitive nature of evolution because of the harmony he felt it had produced. Both inabilities to come to grips with the paradox of evolution reflect real cultural biases. Imanishi's views could and should, like Kropotkin's, have been presented as a special emphasis within the larger evolutionary framework rather than as a brand-new theory.

Explanations for this largely unnecessary quarrel range from that it was the work of journalists, who loved to present one of their countrymen as having the same stature as the great Darwin, to the nationalism of Imanishi himself. Like many

Japanese after the war, Imanishi resented the Anglo-American dominance that made it seem that their own culture didn't matter. In his defense, older scientists stress that Imanishi distanced himself equally from the Russian influence on Kyoto intellectuals, many of whom were left-wing and enamored with Lysenko, the dangerously ideological geneticist under Stalin. Imanishi was first of all a scholar, open to ideas, always instilling independent-mindedness in his students. He himself read everything he could put his hands on, and urged his followers to be international, and to read and write in English.

The stifling impact on primatology and ecology that Halstead referred to did not come directly from Imanishi himself, who retired from the university in 1965, but from his followers. Japanese scientists who embraced sociobiology in the late 1970s or 1980s carry the scars of an uphill battle in a strongly conformist society. In the same way that every silverback gorilla inescapably reaches the point at which he can no longer defend his group and will be overtaken by an upstart, Imanishi and his followers kept looking impressive and in control while under their feet the theoretical landscape was shifting like quicksand. The rebels eventually won, as is evident from the current crop of young primatologists and ecologists, many of whom are either unaware of Imanishi or consider him irrelevant.

Darwin Envy

The anti-Darwinian attitude of Imanishi and his descendants considerably slowed the adoption of sociobiology in Japan.

Partly because of a lack of education in evolutionary theory, Imanishi's made-in-Japan evolutionism prevailed for a long time. Osamu Sakura, an ex-primatologist at Yokohama National University who now studies the reception of sociobiology around the world, has compared the response in Japan with that in France, which was also tinged with national pride and Darwin envy (given that Darwinism replaced Lamarckism). But he found the strongest parallels with the reception of sociobiology in Germany.[86]

When an influential founder, or guru, is on the local scene, new scientific developments may undermine his authority and that of his school. Hence, the new development will be resisted by members of the school. Only outsiders or a younger generation less committed to the guru will dare to venture outside established doctrine. In Germany, this process was clearly visible around Lorenz, whose followers stood in the way of new scientific developments. Thus, both Sakura and Hiraiwa-Hasegawa insist that Imanishi's influence should not be looked at as uniquely Japanese. As Hiraiwa-Hasegawa notes:

> If you assume that the Japanese scientists fully understood the neo-Darwinian theory and still preferred Imanishi's theory to Darwinism because of their original, cultural view to see the organic world, you will end up overestimating a cultural effect on logical thinking in natural science. I think this is what Pamela Asquith did in her research. Or you will end up underestimating the Japanese ability of

logical thinking. I think this is what Beverly Halstead did in his *Nature* paper on Imanishiism.

Her reference to Asquith concerns a Canadian anthropologist who has extensively compared Eastern and Western primatology. Asquith has pointed out many useful distinctions, such as the aforementioned lack of human-animal dualism in Eastern religion. She has also tried to place Imanishi's sociological approach in the context of a culture that ties individual identity to group identity. Asquith doesn't actually believe that Darwin was ever rejected by Japanese scientists, but rather that Western scientists tended toward either-or thinking (they were either for or against sociobiology), whereas their Japanese counterparts simply adopted the most appealing elements from the theory, ignoring the rest.

This leaves us with one unresolved issue. Some believe that Japanese scientists responded the way they did to sociobiology due to a special set of circumstances related to the end of the war, nationalism, and the way founding fathers first stimulate and then hamper progress. In this view, there was nothing specifically Japanese about what happened. Others tie Imanishi's theory to his background and hence see the support for Imanishi and the resistance to sociobiology as stemming from cultural differences. Imanishiism may never have represented a true alternative to Darwinism, but in the eyes of many Japanese it contained valuable, unique elements thus far ignored by the West.

My own position—number one thousand and one—is that Imanishi deserves immense gratitude from all students of long-lived animals in the field since he planted the seed for the only sensible approach. He did so with a typically Eastern emphasis on the group as well as respect for individual identities. At the same time, he downplayed or misunderstood the power of the Darwinian framework, and in doing so blocked much-needed change. But whereas he and his followers had the political power to slow the acceptance of sociobiology, their ideas simply couldn't stem the flow.

It only goes to show that when the time comes, everyone will change.

3

Bonobos and Fig Leaves

Primate Hippies in a Puritan Landscape

"The bonobo is an extraordinarily sensitive, gentle creature, far removed from the demoniacal Urkraft [primitive force] of the adult chimpanzee."
Eduard Tratz and Heinz Heck, 1954

". . . in kinship with our insouciant, fun-loving, nonreading relatives the delightful cousins Bonobo. May Life be thanked for them."
Alice Walker, 1998

"Of bonobos, perhaps the less said the better, at least in a family magazine."
Barbara Ehrenreich, 1999

*I*t must be fun studying those bonobos," people sometimes say to me with a wink. It most certainly is, but not for the implied reason—that this species of ape engages in an aston-

ishing variety of sexual activity. What actually fascinates me
more is the puzzle of how bonobo society came to be female-
centered and pacific. The answer has implications for the evolution of human sexuality and gender relations: every program
in women's studies ought to include a little excursion into the
world of the bonobo.

But the question about their sex life is on target when it
comes to the public reception of the bonobo. The way animals are perceived mirrors the culture in which we live. Thus,
we admire the workaholic ant but not the lazy pig, which in
addition is seen as dirty. Similarly, bonobos have reached us
with a double handicap: first, they seem sexually licentious,
and second, they refuse to fit macho evolutionary scenarios
that revolve around the inevitability of violence, male dominance, male bonding, and the importance of technology. The
traditional outlook has been amply summarized under titles
such as *Man the Hunter, Man the Tool-Maker, Demonic
Males, Men in Groups, The Imperial Animal,* and *The Dark
Side of Man.* These views have become so ingrained that anyone who brings up the sexy and peaceful bonobo risks being
called a dreamer.

In contrast to the pig, however, the bonobo is not some distant life form that can be pushed aside in debates of human
nature. This anthropoid is indeed so close to us that some scientists consider it the best extant model of the last common
ancestor of humans and apes, thought to have lived about five
million years ago. Yet, due to its relatively late discovery (in

1929) and its rarity in captivity, the species was until recently barely known. Nevertheless, bonobos are equally close to us as chimpanzees.

It was not until the 1970s that expeditions by Japanese and Western scientists to the Democratic Republic of the Congo (formerly called Zaire) began to document the natural history of the elusive bonobo. What they found has put question marks all over the evolutionary map of our species, which is commonly believed to have conquered the planet through warfare and genocide. Other primates, such as baboons and chimpanzees, do seem to lend support to these views, but the lives of bonobos most certainly do not. If chimpanzees are from Mars, bonobos must be from Venus.

Bonobos are egalitarian primates that substitute sex for aggression: they resolve conflict through sexual contact. Females occupy prominent positions in society, and the high points of the bonobo's social life are conflict resolution and sensitivity to others. Whereas it would be unwise to rely exclusively on the bonobo in attempts to reconstruct our evolutionary past, the species does expose the one-sidedness of previous attempts.

Obviously, bonobos would not need conflict resolution if they never had any reasons for conflict. It is the paradox of my work that the study of peacemaking requires conflict and aggression. We researchers have ample time to reflect on this if we are working with a colony of primates that gets along "too well," making it hard to collect the necessary information on how they manage conflict.[87] This has never been a problem with bonobos, though, which are far from perfectly pacific. Bonobos are lively and competitive: they are no mellow pushovers who love each other to death. But they do have a different, highly effective way of regulating rivalries, and as such invite an alternative mode of thinking about our ancestors, one that has thus far not occurred to anyone except a few isolated feminist authors.

The bonobo thus serves as a reminder that people who keep shoving the murderous side of chimpanzees into our faces so as to make the point that humans are "killer apes," as Robert Ardrey called us, have a biased agenda. They have seized upon the chimpanzee with an enthusiasm that doesn't do justice to this species—which is cooperative and sociable most of the time—but that does expose the culturally colored glasses that have thus far kept the bonobo on the sidelines of the human evolution show.

Kamasutra Primates

My own involvement with these apes began in 1978, when I first looked a bonobo in the eyes and immediately noticed how their curious and sensuous temperament differed from that of the emotionally volatile chimpanzee. I later set out to study the world's largest colony of bonobos by spending hundreds of hours with a video camera in front of an enclosure at the San Diego Zoo. Work with animals in captivity cannot replace observations in the field, but it does offer the enormous advantage of complete visibility so that behavior can be followed in its most minute details. I was thoroughly familiar with chimpanzee behavior, which I had interpreted in rather Machiavellian terms in *Chimpanzee Politics*. Now, I was seeing a tableau more reminiscent of Rousseau—or perhaps I should say a simian version of the Kamasutra.

Bonobos have sex in all imaginable positions and in virtually all partner combinations. They contradict the notion that sex is intended solely for procreation. I estimate that three quarters of the sexual activity I saw at the zoo had nothing to do with reproduction: it frequently involved members of the same sex or took place during the infertile portion of a female's menstrual cycle. Sexual activity was most likely to occur at moments of tension, such as when there was a risk of competition over food, or as a way of reconciliation after a fight.

It is impossible, therefore, to understand the social life of the bonobo without attention to its sex life. Whereas in most other species sexual behavior is a fairly distinct category, in the bonobo it has become part and parcel of social relationships. Bonobos become aroused remarkably easily, and express this in a variety of mounting positions and genital contacts. Perhaps the most characteristic sexual pattern is so called genito-genital rubbing, or GG-rubbing, between adult females. One female clings with arms and legs to another—almost the way an infant clings to its mother—while the other female, standing on both hands and feet, lifts her off the ground. The two females then rapidly rub their genital swellings laterally together. Completely absent in the chimpanzee, this behavior has been observed in every bonobo group, captive or wild, with more than one female.

Male bonobos, too, may mount each other, but generally perform a brief scrotal rub instead: standing back-to-back, one male rubs his scrotum against that of another. They also show so-called penis fencing, a rare behavior, thus far observed only in the field, in which two males hang face to face from a branch while rubbing their erect penises together as if crossing swords. The sheer variety of erotic contacts is impressive, especially if we include the sporadic oral sex, massage of another individual's genitals, and tongue-kissing.

The same within-group use of sex seems to extend to relations between groups. This is quite a contrast with chimpanzees, in which males are known to patrol the borders of their territory and occasionally invade their neighbors', setting off lethal battles. In bonobos, there is not a single report of this level of intercommunity violence. Instead, peaceable mingling of communities seems to be the rule, including mutual sex and grooming.

The Two Laws of Puritanism

The bonobo's sexiness has been a mixed blessing for its public reception. The problem first emerged shortly after the last world war, when the same German scientists who came up with the bonobo's unusual name felt a need to explain that this ape often mates face to face. In those days, this was an unmentionable detail. Eduard Tratz and Heinz Heck had to resort to Latin, saying that chimpanzees mate *more canum* (like dogs) and bonobos *more hominum* (like people). They added that female bonobos are anatomically adapted for this position: the vulva is situated between her legs rather than oriented to her back, as in the chimpanzee.

Nowadays, in the same Munich where Tratz and Heck made their pioneering observations, one can watch young professionals from downtown enter the Englischer Garten for a lunch break, sit down in the grass, take off their clothes, neatly fold them by their side, and continue their conversations in the nude. Because they are not doing anything else than what happens daily on European beaches, no one blinks an eye. Continental European attitudes have indeed changed radically since the 1950s, making them quite different from those in my adoptive country.

I realize that to complain about "Americans" is a not an altogether endearing European pastime, but it is impossible to discuss bonobos without a word about puritanism. Although I have lived in the United States for two decades, and have a genuine fondness for the country and its people, I will never get used to the equation between sex and sin. The guilt and suffering—not to mention the hypocrisy—this association creates are beyond me. I would gladly avoid this topic if it weren't for the question that persistently arises when people first hear of bonobos—namely, why this species isn't more widely known. The answer is, at least in part, that they remind us too much of a side of ourselves that we fight hard to control. Instead of being hardworking and chaste they lead promiscuous, hedonic lives. If they are our closest relatives, better keep them locked away!

Now, I know many Americans who are quite open-minded about sexual matters, but unfortunately their society is not. I will call this the First Law of Puritanism: the whole is more puritanical than its parts. The tension between public morals and private thoughts tends to be overlooked by public servants and the media, who customarily err on the conservative side when judging the acceptability of behavior or materials. Thus, they may try to hang a public figure for his transgressions only to discover that the majority of people feel that a mere reprimand would do.[88]

The Second Law is that sexual repression is harder to see from the inside than the outside. Americans are so used to living in a country where toilets are called restrooms, where even gynecologists don't see naked patients, where one can get arrested for breast-feeding in public, where pinups come in swimsuits, and where comedians shock audiences into convulsive laughter by merely dropping the name of a taboo body part, that they don't realize how peculiar all of this looks from the outside.[89] A possible exception are Americans who have traveled abroad, where they may have visited a Japanese bathhouse in which removal of all clothes is obligatory even in the presence of the opposite sex. They may also have seen free and open prostitution in Amsterdam and Hamburg, or met people who simply shrug their shoulders at the sex lives of their leaders.

The most recent example where I felt out of touch with American attitudes was a 1999 cover story in *Time* magazine, bravely entitled "The Real Truth About the Female Body."[90] To illustrate this truth, *Time* offered us one cover photo plus five photos inside the magazine of nude women. It managed to do so without revealing a single nipple or genital area. The bodies shown were muscular and androgynous: one had to look hard to make sure they were female. The magazine went so far as to include a foldout body with arrows pointing at its various parts, but since the woman in the photograph held her arms crossed in front of her, the arrow intended for her breasts sadly hit an elbow.

Because it deliberately expunged all femininity from the female body, this layout would have provoked screams of protest if featured in an equivalent European magazine, such as *der Spiegel* or *l'Express*. *Time* got its share of protest, too, but this was because by American standards they had gone too far! In a subsequent issue, the editors noted, "Many readers felt that nude pictures of women, however tastefully done, have no place in a general-interest magazine."[91]

And so, in sexual matters, the two North Atlantic continents have drifted apart, even though they share in many ways the same background. Together they differ from non-Western peoples, such as the Hawaiians, for whom sex has been described—in rather bonobolike terms—as "the salve and glue for the total society,"[92] and from some Brazilian tribes, where men and women carry on multiple affairs in the woods around their villages. This is not the place to review the thousands of human sexual practices; suffice it to say that on a global scale of permissiveness and openness much of the English-speaking world, with its Victorian heritage, occupies a position at the squeamish end.

Given that the English language dominates the information flow in the modern world, this has not been to the advantage of the bonobo. When the Make Love Not War hippies of the animal kingdom knocked at our door, they were left standing outside by a mortified family. The author of the same *Time* article about the female body, Barbara Ehrenreich, felt that the bonobo's peculiarities were better left alone. Similarly, a British camera crew traveled all the way to the remote jungles of Africa to film bonobos only to stop their cameras each time an embarrassing scene came into view.

The crew was assisted by Takeshi Furuichi, a Japanese scientist extensively familiar with the role of sex in bonobo society. When Furuichi politely inquired why they did not document any of it, the answer was "our viewers would not be interested."

What's Wrong with Those Males?

The beauty of sex is that however much a society tries to suppress it, it won't succeed. Sex always bounces back. People will

keep doing what nature has instructed them to do, no matter
how many Sodom and Gomorrah sermons they receive. They
know their own weaknesses, and moreover they have not failed
to notice how the great moralizers of society do not always live
up to their own standards. Hence the First Law of Puritanism:
what society as a whole denounces may well be acceptable, or
at least forgivable, in the hearts of most individual members.

The bonobo is a case in point. Not only are some minori-
ties, such as homosexuals and polyamorists, for obvious reasons fascinated by these lusty apes, mainstream America
seems to have embraced them as well. It turns out that the initial reluctance of the media reflected a misjudgment of public
sensitivity: bonobos appear rather less shocking and offensive
than anticipated. I remember telling U.S. television producers
that I had shown uncensored bonobo footage on Italian,
German, and Dutch prime time television. Why can't the
same be done here? Some producers would meet the challenge, promising that they could do the same on American
television. At the last minute, however, they invariably chick-
ened out. The documentary would show frolicking bonobos
but freeze the image as soon as they adopted positions in
which something sexual was imminent. The narrator would
lead the viewers astray with some vague statement, such as
that bonobos are remarkably friendly with each other. I began
to dub it the *coitus interruptus* treatment.

After years of this, I had the luck of running into a kindred
spirit who had gone through similar frustrations. One day in

1994, Frans Lanting, a celebrated wildlife photographer, told me about the hundreds of pictures of bonobos he had taken during a National Geographic expedition to the Congo. Most of them had never seen the light of day for reasons that by now should be obvious. When I saw the treasure trove of wonderful shots taken under the most trying circumstances (there is nothing worse for a photographer than black animals in a dark forest), I immediately realized that the pictures presented a momentous opportunity. As Dutchmen of the same age living in America, Frans and I had a quick rapport, and we decided to work together on a book about the bonobo to raise public awareness of this special ape.

The goal was to tell the full story. In our minds, this didn't necessarily imply emphasis on sex and eroticism, because there is much more to bonobos than that. But it did mean that we would not let ourselves be censored. Thus, the first report we put together was for a well-known, richly illustrated German magazine. GEO had no trouble printing an undiluted account, with erect penises, pink genital swellings, "homosexual" acts, and all.[93] The next trial balloon was Scientific American, which to its credit didn't change a word—except for stylistic reasons—of my text, and reproduced all of Lanting's photos. By then, we felt the time was ripe for a book. We found an American university press prepared to accept our condition that no censure would be applied. Perhaps not accidentally, the publisher was located in Berkeley. They held their word, resulting in Bonobo: The Forgotten Ape (1997), which reached a large audience without provoking a single peep of moral outrage that I know of.

In writing this book, I saw it as my first task to draw in as many bonobo experts as I could. The peacefulness of our research subjects notwithstanding, our little field has not been without the usual infighting among close scientific colleagues. I wanted none of that reflected in the book. So, I conducted interviews with many of my colleagues so that they could say what they had to say in their own words. This way, I also hoped to avoid giving the impression that I had discovered everything on my own. I am not a field-worker, and researchers such as Takayoshi Kano, the Japanese scientist who for twenty-five years operated a field site under the most trying conditions, deserve a great deal of credit for what we know today about bonobos.

A sabbatical in Europe allowed me to devote my time to writing. I alternated my hours behind the computer with train trips around Austria, Germany, and the Netherlands to test my message on all sorts of audiences. The high point of my European lecture tour—or perhaps it was the low point—occurred when an older, highly respected German professor stood up after my lecture and barked in an almost accusatory tone: "What's wrong with those males?!" He was shocked by the dominance of females. Given that bonobos thrived for thousands of years in the African rain forest until human activity began to threaten their existence, there really seems nothing wrong with them at all. And in view of their frequent sexual activity and low aggression, I find it hard to imagine that males of the species have a particularly stressful time. My response to the professor—that bonobo males seemed to be doing fine—did not appear to satisfy him. The incident, though, shows how profoundly the bonobo is challenging assumptions about our lineage.

Uncomfortable Scientists

I received the first hint of this ape's unconventional social order when I revisited the San Diego Zoo one year after my initial study. Originally, Vernon, an adult male, had been housed alone with Loretta, an adult female whom he had dominated. But when I returned, Louise, an older female, had been added to the group, and she and Loretta now clearly bossed over Vernon. In fact, Vernon had to beg the females to share food with him, and Louise sometimes chased him. I found this odd, since Vernon was a muscular male who not only was larger than Louise or Loretta but also possessed the sharper canine teeth of his sex. As I came to know more captive groups of bonobos, however, I found that female dominance was the rule rather than the exception.

Field-workers suspected the same for an even longer time. Yet bonobo specialists have been reluctant to make such a controversial claim—until 1992, that is. At the XIVth Congress of the International Primatological Society, in Strasbourg (France), investigators of both captive and wild bonobos for the first time presented data that left little doubt about the issue. Amy Parish, an anthropologist from the University of California at Davis, reported on food competition in identical groups (one adult male, two adult females) of chimpanzees and bonobos at the Stuttgart Zoo. Honey was provided in a place from which it could be extracted by dipping sticks into a small hole. As soon as honey was given, the male chimpanzee would make a charging display and simply claim everything for himself; only when his appetite was satisfied would he let the females fish for honey. In the bonobo group, in contrast, it was the females who approached the honey first. After some GG-rubbing between them, they would feed together, taking turns with virtually no competition. The male could make as many charging displays as he wanted; the females were not intimidated, and ignored the commotion.[94]

At the same conference, field-workers confirmed that female bonobos dominate males. For example, at the provisioning site of Wamba in the Congo, males always arrive first and eat in a hurry, because when the females come they are forced to make room. Some scientists have questioned whether this pattern should be called dominance, proposing instead that male bonobos are tolerant and deferential. It's almost amusing: males are usually painted as competitive monsters, yet if they systematically lose battles with females it must be because they are nice guys.

However, if there is one criterion that we have used for every animal on the planet, it is that if individual A can chase

B away from its food, A must be dominant. It is unclear why we should suddenly adopt a different standard for bonobos. Kano has forcefully rejected this argument:

> Priority of access to food is an important function of dominance. Since most dominance interactions, and virtually all agonistic episodes [conflicts] between adult females and males occur in feeding contexts, I find much less meaning in dominance occurring in the non-feeding context. Moreover, there is no difference between feeding and non-feeding dominance relationships among the bonobos of Wamba. For example, approaches of dominant females often give rise to submissive reactions by grooming males such as grinning, bending away, etc.[95]

In the same way that investigators looked through cultural lenses at the bonobo's social organization, they have been unable to shake the prevailing moral biases when it comes to sex. Some went so far as to question the "sex" label for contacts between same-sex partners. True, social goals are often accomplished by such contacts, such as when dominance or affection is demonstrated in a sexual fashion. Yet they are still sex. In common language, sex covers all deliberate contacts involving the genitals (including petting and oral stimulation). It even includes broader categories, such as kissing or just showing off one's body in a suggestive manner. More than once, however, I have heard that GG-rubbing among bonobos doesn't deserve to be called sex—perhaps it is nothing more than mutual masturbation. Such an argument, however, overlooks the intensely social and apparently enjoyable nature of the interaction: when females rub their prominent clitorises together, they often bare their teeth in a wide grin, uttering excited squeals while they look each other in the eyes. If this is mutual masturbation, shouldn't heterosexual intercourse be reclassified as well?

In his book *Biological Exuberance*, Bruce Bagemihl cites a great many instances of what he sees as homophobia in the scientific literature. Routinely, adjectives such as "pseudo" and "sham" are added to labels for sex between partners of the same sex (as in "sham sex" or "pseudo-copulation"). Authors of otherwise serious papers sometimes express disappointment in a species that shows such "disgusting" behavior, or else a journal editor adds a footnote that provides an alternative, nonsexual elucidation. Perhaps the most imaginative speculation of this kind was one that attributed a "nutritive" motivation to two orangutan males who regularly sucked each other's penises. Bagemihl notes:

> When a male Giraffe sniffs a female's rear end—without any mounting, erection, penetration, or ejaculation—he is described as sexually interested in her and his behavior is classified as primarily, if not exclusively, sexual. Yet when a male Giraffe sniffs another male's genitals, mounts him with an erect penis, and ejaculates—then he is engaging in aggressive or dominance behavior, and his actions are

considered to be, at most, only secondarily or superficially sexual.[96]

It is in this light that we may understand why an American primatologist, Craig Stanford, concluded in a recent critique that bonobos aren't any more sexual than chimpanzees. Perhaps due to lack of experience with bonobos, Stanford restricted his calculations of mating frequencies to heterosexual contacts, leaving out an enormous portion of the species' sex life. He also surmised that since the most detailed descriptions of bonobo sexuality come from zoo environments, it might all be an artifact of captivity. Perhaps these apes act so grotesquely because they are bored to death, or under human influence. However, under identical zoo conditions bonobos and chimpanzees act totally unlike each other. If captivity distorts the behavior of one ape, why not also that of the other? The inescapable conclusion is that it is something in the species, not the environment, that produces the bonobo's characteristic sexuality.[97]

Field data support this inference. In Lomako Forest, where bonobos are studied without human food provisioning, sexual activity increases during the sharing of meat or when a group of bonobos excitedly enters a fruiting tree. And at Wamba, sexual activity in all partner combinations is common when there is a potential for competition, such as when the investigators provide sugar cane. In short, there is no relevant discrepancy between how bonobos behave in captivity and in the field.

For example, at the zoo bonobos sometimes resolve competition through food-for-sex exchanges. Similarly, Suehisa Kuroda saw in Wamba that "a young female approached a male, who was eating sugar cane. They copulated in short order, whereupon she took one of the two canes held by him and left." In another case, "a young female persistently presented to a male possessor, who ignored her at first, but then copulated with her and shared his sugar cane."[98]

Such transactions provide, I believe, a fascinating window on the bonobo's past. Most likely, the species started out similarly to the other apes, that is, with male dominance. In the course of evolution, sexual receptivity may have been extended, resulting in longer-lasting genital swellings that helped females bargain for food controlled by males. Over time, this tactic became more and more restricted to young females. Fully adult females tend to be of equal or higher status than males, and simply claim food whenever they want.

Female bonobos establish close ties amongst themselves, and mothers exert such influence on the lives of their sons, including fully grown ones, that Kano has called mothers the "core" of bonobo society. For example, females meddle in fights among males, and in doing so determine which males will be high-ranking. Although I certainly don't buy the feminist myth of hominid ancestors free from gender biases, bonobo society does come close to what the writer Marilyn French, in *Beyond Power*, labeled a "matricentry."

New Kid on the Block

Chimpanzee experts are spoiled by the quirk of history that made the species they study known long before the bonobo. They have become so used to dropping the catchphrase "man's closest relative" in connection to their subjects that they have a hard time getting used to the modifier "one of man's closest relatives" in light of the bonobo, which is equally close to us. Moreover, the bonobo's sexiness has, after a period of discomfort, turned these primates into media stars. They are not merely sharing the limelight, they have begun to steal it![99]

At a more profound level, the bonobo's female-centered society is inconvenient for those who are invested in male-biased evolutionary scenarios. Whereas the chimpanzee perfectly fits this line of thinking, the peaceful bonobo is urging a reconsideration of the underlying assumptions. As a result, the handful of scientists familiar with these apes have been put in the position of defending their observations against skeptics who mostly have never laid eyes on a bonobo. Since they cannot make the new kid on the block go away, they question how special it really is. As a result, we are seeing the implausible spectacle of serious scholars denying conspicuous bonobo features and coming up with imaginative alternative explanations. They try to tell us that certain kinds of sex are not sex, that female dominance might be male chivalry, and that female bonding could be mere tolerance.[100]

As one of the few scientists familiar with both bonobos and chimpanzees, and a firm believer in the profoundness of their differences, I had a special question for Furuichi and his wife, Chie Hashimoto, when I met with them at a Yokohama bar over a giant tuna head from which we picked meat with chopsticks. Due to the Congo's political upheavals, the couple has given up on bonobo fieldwork, and now studies chimpanzees in Uganda. Here I had two of the very few people in the world who intimately know both ape species in the field. "Do you feel the differences between bonobos and chimpanzees have been exaggerated?" I asked. They almost jumped off their stools, exclaiming how *shocked* they had been by chimpanzees. They had seen with their own eyes what Tratz and Heck called the "demoniacal *Urkraft*" of the chimpanzee, its stormy temperament, its brutal competitiveness, but also its male bonding and unique political complexity. Before going to Uganda, they themselves had suspected that perhaps the literature had painted too much of a black and white picture, but they were now convinced that bonobos and chimpanzees lived in truly different worlds.

Kuroda, who also has experience with both species in the field, told me a telling detail about how bonobos flee from strange people. Whereas chimpanzees scatter in all directions, bonobos stay together as a group when they escape unwanted attention. In the chimpanzee, even mother and young may take separate routes, which Kuroda, who started out on bonobos, found shocking, since these apes would never do such a thing. It is also known that bonobos call each other so as to congregate before they build nests for the night, whereas chimpanzees usually sleep on their own. The two species' temperaments seem radically different, with chimpanzees being independent-minded, and bonobos highly sociable and solidary.

As soon as camera crews are able to enter the Congo again, and have the guts to film bonobos the way they are, people will understand that everything being said about these apes is no exaggeration. They are not the product of some overactive sexual imagination, or of wishful thinking. That they delight feminists, the gay community, and pacifists should not be held against them. If one of our closest relatives fails to fit the prevailing views about aggressive males and passive females, one possibility to consider is that the prevailing views are mistaken.

Either that, or there is something wrong with those males.

4

Animal Art

Would You Hang a Congo on the Wall?

"The ape would try new ideas, make new patterns, but only very slowly. Old patterns were repeated and only slightly altered as time went by. The war between wildness and security, between strangeness and familiarity, these were being worked out by the ape, at its simple level, just as human artists were working them out at their highly complex and advanced level."

Desmond Morris, 1997[101]

Whehn advanced paintings and engravings were discovered on rocks in South Africa, the first reaction of Western experts was that these creations could not possibly have come from the indigenous San, or Bushmen, as this would mean that the San had discovered the power of art on their own. The actual artists must have come from the outside.

In line with this view, Henri Breuil, the great French expert on Upper Paleolithic art, named a Namibian painted figure *The White Lady of Brandberg* because he could tell her racial origin and felt she must have been Mediterranean. Others believed that prehistoric Europeans, navigating around Africa in search of hunting grounds, had produced the paintings. But after closer examination, the evidence is now overwhelming that the San were responsible for the rock art, that Breuil's white lady was neither Caucasian nor female, and that some of the art is older than the famous Lascaux cave paintings in France.[102]

The reaction to the San rock art is emblematic. It reminds one of the reception of the first European cave paintings. In 1879, a little girl and her father looked at the ceiling of a low cave in Altimira, Spain, and saw dozens of bison, horses, boars, deer, and a wolf in the flickering light of their oil lamp. The father, an amateur archeologist, reported the find, but did not live to see it accepted as genuine prehistoric art, which happened only decades later. Initially, the cave paintings were dismissed as the product of modern artists. It was just not conceivable that primitive minds could have produced images of such elegance, realism, and artistic beauty.[103]

Underlying the skepticism surrounding these and other early finds is the idea that it is only recently, and only in very few human populations, that cultural sophistication has reached a level permitting artistic images. Art is supposed to set civilized man apart from the rest. It is regarded as even more characteristically human than language and culture, a capacity we are extremely reluctant to grant even to primitive folk. If the trait is that exclusive, animals obviously deserve no mention at all.

Yet biologists feel that animals are no strangers to aesthetic expression. The New Guinean bowerbird's nest decorations are as good an example as any. The thatched nests can be so large and well-constructed that they once were mistaken for the huts of timid people, who never showed up. The nests often have a doorway with carefully arranged colorful objects, such as berries, flowers, or iridescent beetle wings. The male who built the bower keeps flying in new ornaments, shifting everything around with a critical eye, fussing over the arrangement, moving back to look at the whole from a distant angle — like a human painter with his painting — and then continuing the rearrangement. He is very sensitive to the fading of his flowers, replacing them with fresh ones as soon as necessary. Young males build crude "practice" bowers, tearing them down, then starting over again, until the construction holds up as it should. They also frequently visit the completed bowers of adult males in the neighborhood and see how the ornaments are laid out. There are ample learning opportunities here, and it has been noted that bower decorations differ in color and arrangement from region to region, which suggests culturally transmitted styles.[104]

Is this art? One could counter that it isn't: bowerbird males are genetically programmed to engage in this activity just to attract females. Yet, while it is true that females select mates on nest quality and their equivalent of a stamp collection, the argument is not nearly as good as it sounds. To contrast these birds with our species requires that one demonstrates that human art does not rest on an inborn aesthetic sense and is produced purely for its own sake, not to impress anyone else. Both are unlikely. In fact, Geoffrey Miller argues in a recent book that impressing others, especially members of the opposite sex, may be the whole point of human art![105]

What if our artistic impulse is ancient, antedating modern humanity, and perhaps even our species? What if it rests on a delight in self-created visual effects and a penchant for certain color combinations, shapes, and visual equilibriums that we share with other animals? Would admission in any of these areas diminish the significance of and pleasure derived from human art? Isn't it possible that our basic distinctions in art, our musical scales, and our preference for symmetrical compositions, go deeper than culture, and relate to basic features of our perceptual systems?

What better way to connect humans and animals culturally than investigate the common ground in the visual arts and music? There obviously remain vast differences, but from an evolutionary perspective it would be strange indeed if the beauty that we recognize in nature, and that has inspired so many human artists over the ages, would have an impact only on our own species. Our eyes and ears are very similar to those of many other life forms, and until very recently we dwelled in the same kind of environment. The ancestral environment must have shaped our senses, making us seek certain impressions more than others. This argument has regularly been made in architecture, such as the claim that the famous *Court of Lions* in the Alhambra, in Granada, evokes a universal emotional response because standing inside the columned arcade, looking at the lighter part in the center, harks back to walks through the forest looking out at open areas. It is easy to agree, therefore, with the following recommendation by Nicholas Humphrey: "If I were asked for a prescription for where architects and planners should go to learn their trade, it would be this: Go out to nature and learn from experience what natural structures men find beautiful, because it is among those structures that men's aesthetic sensitivity evolved."[106]

Can't Stand Schönberg

Everyone with an ear for music appreciates the moving, pleasing quality of birdsong, especially that of species with variable, long-phrased repertoires such as the nightingale or blackbird (a European relative of the American robin). In the time before radio and television, this was the sort of "music" most often heard in the evening, and was treasured by poets and romantic lovers alike. As the twelfth-century Marie de France most famously put it in her poem *Laüstic*, anyone who hasn't heard the nightingale sing doesn't know the joys of the world.

Conversely, animals can be quite sensitive to human music. There are stories of dogs who hide under the couch for piano works by atonal composers but not for those by, say, Mozart. One music teacher told me that her dog would heave an audible sigh of relief if she stopped playing complex, fast-moving pieces by Franz Liszt and proceeded to something calmer. And there are reports of cows that produce more milk listening to Beethoven (although, if this is true, shouldn't one hear more classical music on farms?).

In the laboratory, sparrows have been tested on their preference for composers. Out of four birds, two liked to sit on a perch that turned on Bach rather than perches associated with the twelve-tone music of Schönberg, or white noise. The choices of the other two birds were less clear. Did I detect some hidden glee when the investigators dryly concluded that Schönberg may possess some "aversive stimulus properties"?[107]

Birds listen as carefully to sounds as any musician. They have to, because they learn from each other. Many birds are not born with the song they sing: the symphonies they offer us for free in forests and meadows are cultural. White-crowned sparrows, for example, develop their normal song only when they have been exposed early in life to the sounds of an adult of their species. Many songbirds have *dialects*—differences in song structure from one population to another. One theory about this is that if a female can tell from a male's song that he is a local boy, she may prefer him as a mate, as he may be genetically adapted to regional conditions. Given the variability in song from location to location it is hard to maintain that birdsong is instinctive in the usual sense. There is room for creativity and modification. Some individuals act as star performers, setting new trends in their region.[108]

Inasmuch as birdsong is shaped by oral tradition, this is a potential area of crossover between animal and human culture. A composer may be inspired by what he or she hears in nature, and translate a bird's vocal innovations into a human cultural medium. At a recent meeting, the late Luis Baptista of the California Academy of Sciences reviewed the evidence in a delightful lecture, *Why Birdsong Is Sometimes Like Music*, full of comparisons between the West's great composers and the smaller feathered ones studied by Baptista and his ornithological colleagues.[109]

That composers have often found inspiration in nature is reflected in the titles of their works, such as *Der Wachtelschlag* (The Quail Song)—a title used by three different composers: Beethoven, Schubert, and Haydn—as well as Vivaldi's *Il Gardinellino* (The Goldfinch) and Mozart's *Spatzenmesse* (Sparrow Mass). Bird sounds can be discovered in many works, such as chickadees in Bruckner, pigeons in Britten, and nightingales in Respighi. The most popular bird may be the cuckoo, the unmistakable call of which can be heard in many works, from Beethoven's *Pastorale* to J. S. Bach's fugue *Thema all' Imitatio Gallina Cucci*, which pitches a cuckoo in counterpoint against a chicken, and lets the first win.

Birdsong often follows a sonatalike structure in that it starts out with a theme, followed by variations on it, after which the original theme is recapitulated. The similarity is no accident given that both people and birds get bored by repetitions, and hence need to break the monotony without losing sight of the unifying theme in their compositions. Successful themes are handed down for many generations. Thus, the rondo in Beethoven's Violin Concerto in D Major, which premiered in 1806, features a melody that was independently recognized as the song of a blackbird by both a British ornithologist, in 1953, and a German concert pianist, in 1980. This could mean that Beethoven was inspired by a blackbird-invented melody that these songbirds have kept going, through transgenerational imitation, for over a century.

But the most striking and amusing case is that of a Mozart composition that has baffled musicologists ever since it appeared in 1787.

Mozart's Little Fool

Music historians have found it hard to accept that one of the most idolized Western composers, Wolfgang Amadeus Mozart, could have arranged a solemn ceremony, with veiled, hymn-singing mourners, and a special poem by the composer himself, for the burial of a mere bird. Could it be that, since Mozart's father had died in the same week, the funeral was related to this family tragedy instead? This conjecture hardly explains, though, why on this sad occasion, on June 4, 1787, the great composer's recital began with these lines:

A starling bird rests here,
a fool whom I held dear.
Who in his prime still,
swallowed death's bitter pill.[110]

Anyone familiar with the European starling, *Sturnus vulgaris*, knows how apt this description is (the German word for "fool" in Mozart's poem, *Narr*, also means "jester" or "clown"). The same ordinary bird is now common in the United States because a different kind of fools released over one hundred of them in New York's Central Park in the 1890s as part of an effort to introduce the entire avian cast of the Shakespearean theater. With several hundred million starlings now blackening the skies across the North American continent, the amount of agricultural havoc created by this well-intended decision has been immeasurable.

Starlings *are* clowns, and no one knows this better than the people who have raised these overactive birds at home. They imitate all sorts of sounds made by other animals, people, and objects, such as telephones, rattling keys, and clinking dishes. In the households of academics, they have been known to pick up phrases, such as "basic research" and "I think you're right," which they use at inopportune moments, resulting in amusing commentaries. One bird had a custom of landing on a shoulder while uttering "Basic research, it's true, I guess that's right." Another bird, squirming while being held for treatment of its feet, screeched "I have a question!"

27. May 1784 Vogel Stahrl 34 Kr.

Das war schön!

Piano Concerto no. 17 in G Major, K. 453

Mozart's starling whistled a tune on May 27th, 1784, that thrilled Mozart (top). With a minor modification, the final movement in Mozart's pianoconcerto No. 17 features the same theme. (From Nottebohm, 1880).

Two American birdsong experts, the wife-and-husband team Meredith West and Andrew King, explain the joy of keeping pet starlings at length in an article entitled "Mozart's Starling," (1990). They also provide an analysis of how these birds flexibly combine and recombine song phrases, adding whistles and typical starling squeals to tunes that were once sung to them. They fracture the phrases, sing them off-key, and delete parts that seem absolutely critical to the human ear. For example, one bird would whistle the notes corresponding to "Way Down Upon the Swa-," never, despite thousands of promptings, adding the notes to "-nee River."

In describing the peculiarities of starling mimetics, the authors try to throw light on Mozart's fascination with his bird. He entered the purchase of his starling in his diary and added the transcription of a song it whistled, commenting *Das War Schön!* ("that was beautiful"). It was a familiar tune, almost identical to a theme in the final movement of Mozart's Piano Concerto in G Major. But how could the bird have sung this tune on the date it was bought, May 27, 1784, when Mozart had catalogued his concerto as finished on April 12 of the same year? Speculations hinge on the possibility that Mozart, like many animal lovers, had visited the pet shop in the weeks preceding the purchase, and had transmitted the tune to the bird. The composer was known to whistle a lot, and starlings don't need to hear a melody many times to copy it. Who knows, the composer may have bought the bird out of delight over its mimicry.

Others have speculated about transmission in the opposite direction, that is, from the bird to the composer. To some Mozart fans this may sound sacrilegious, but the alternative is perhaps even worse: independent genius!

With ears trained quite differently from those of musicologists, West and King also listened to Mozart's *A Musical Joke*, the first piece he wrote following the death of both his father and the starling. This piece is commonly interpreted as a parody of the popular music of Mozart's day, or else as a commentary on the father-son relationship. But Mozart's relationship with his father surely didn't deserve this kind of mocking commemoration. Instead, the two bird experts note the piece's starling-like qualities. Consider the following description of K. 522 from a record jacket:

In the first movement we hear the awkward, unpropor-
tioned, and illogical piecing together of uninspired material.... [Later] the andante cantabile contains a grotesque
cadenza which goes on far too long and pretentiously and
ends with a comical deep pizzicato note . . . and by the
concluding presto, our "amateur composer" has lost all
control of his incongruous mixture.[111]

West and King comment:

Is the piece a musical joke? Perhaps. Does it bear the vocal
autograph of a starling? To our ears, yes. The "illogical
piecing together" is in keeping with the starlings' inter-
twining of whistle tunes. The "awkwardness" could be due
to the starlings' tendencies to whistle off-key or to fracture
musical phrases at unexpected points. The presence of
drawn-out, wandering phrases of uncertain structure also
is characteristic of starling soliloquies. Finally, the abrupt
end, as if the instruments had simply ceased to work, has
the signature of starlings written all over it.[112]

Baptista adds to this analysis by noting the final cadence in
A Musical Joke, which is written in two voices in counterpoint.
Funny? Perhaps, but birds produce sounds with a syrinx that
has two vocal cords, which can act independently, allowing a
two-voice phenomenon that would make Bach proud.
Baptista agrees, therefore, that the composition must have
been Mozart's final farewell to his four-penny bird. In this
light, many of the jargon-laden analyses that I have read be-

come truly amusing. Almost all musicologists assume either that Mozart got lost in his own music (calling this particular composition superficial, and devoid of significance) or that he spoofed contemporary colleagues who had trouble composing. But they all miss the real joke! One Czech colleague, Leopold Kozeluch, is even said to have attacked Mozart on a visit to Prague because he felt parodied.

It has come to light that A *Musical Joke* was composed in fragments during exactly the three-year period that Mozart owned his darling starling. Its completion a week after the bird's death suggests that it was a requiem for his avian friend. People who share Mozart's love for birds (he also kept canaries), and who know the naughty and endearing qualities of the starling, have no trouble believing he felt a great loss. Birds develop strong attachments, showing a tender and happy side to those they love and trust. They may gently nibble at their owner's ear, for example, making soft sounds of contentment, where they might peck someone else's. We people have a natural tendency to reciprocate when we notice how much we mean to another being. In Mozart's case, this special bond was enriched by mutual inspiration between the professional composer and his feathered amateur colleague.

Pigeons and Impressionists

Of all paintings by famous artists on the market and in museums, ten to forty percent are estimated to be fakes—perfectly good paintings, but by different artists than is claimed for them. But with art experts staking their reputations on existing classifications, it is hard to change opinions. When forger Han van Meegeren claimed that he was behind some of the best-known works attributed to Jan Vermeer, no one wanted to believe him. During the German occupation of the Netherlands, he was arrested for selling the enemy a painting by the Dutch master. The only way for him to prove that he himself had produced the art—a lesser offense than collaboration—was to paint one more "Vermeer" while in prison.

This is why we need more pigeons, the only experts unfazed by big names, astronomical prices, and paper authentication. Near a sports field at Japan's oldest and most prestigious university, Keio University in Tokyo, Shigeru Watanabe runs a modest but crowded laboratory in which students and collaborators are constantly placing birds and other animals in test chambers to measure one perceptual ability after another, such as whether pigeons can detect the difference between healthy and sick members of their species, between Schönberg and Bach, or between a Monet and a Picasso. To the astonishment of the art world, which considered discrimination among painters an acquired taste attainable only by one aesthetically sensitive species, Watanabe's pigeons have no trouble with the latter task.

One group of pigeons was rewarded for pecking at pictures of Monet's paintings, and another for pecking at Picasso's. After the training was over, the same birds were presented with new paintings, never seen before, but by the same artists. They generalized from the pictures on which they had been trained to the unfamiliar set. So, a pigeon trained on Picasso's *Girls in Avignon* and *Nude Woman with a Comb* would also peck at *Woman Looking at the Glass* and *Natura Morta Spagnola* by the same artist. Similarly, birds trained on Monet would generalize from one set of paintings by this artist to another. Since we don't assume that the pigeons see two-dimensional images as representations of the real world, it is unlikely that their distinctions were based on objects recognizable to us (women versus fruit, for example). One might therefore conclude that the cue must be the color scheme, together with the presence or absence of sharp edges. However, when Watanabe modified the paintings by presenting them in black and white or with blurred lines, the birds were still able to make the discriminations.

There is more. When the same birds were asked to peck at paintings by other artists of the same period, the Monet-trained birds preferred other impressionists, such as Renoir, whereas the Picasso-trained birds preferred other cubists, such as Braque. So, pigeons can pick out not only individual styles, but entire schools of visual art. Watanabe thinks his pigeons make complex visual distinctions in the same way we do, using multiple cues all at once. The fact is, they distinguish painters better than many a visitor to the Louvre.[113]

But what about the *production* of visual art? Although animal art is on the market, some of it really doesn't qualify because it is randomly produced. There is, for example, the case of the orangutan at a major zoo who would search for a rock, then bang it against the glass wall of his enclosure with such superhuman force that it would shatter and he could escape. Despite the zoo's efforts to remove all rocks, he kept finding them, or digging them up. The wall-shattering became such a predictable event that the zoo paid for its regular purchases of expensive bulletproof glass by setting up a small business. The fractured slabs of glass were successfully sold as orangutan-produced tabletops, making, no doubt, for excellent conversation pieces.

Such unintentional animal "art" is widespread. One of the classic examples is the way the Japanese artist Hokusai won the favor of his shogun, in 1806, by unrolling a lengthy piece of paper on the ground and covering it with big blue loops. He then took a cock, its feet dripping with red paint, and made it walk across the paper. To the Japanese eye, the result looked immediately like a river with floating red maple leaves.

The animal-as-paint-tool was exploited more recently by dipping cats' paws in paint so that they put colorful marks all over the place. This led to a tongue-in-cheek photo book (perhaps to be placed on the orangutan coffee table) that included touching portraits of the artists, complete with personal traumas and van-Gogh-like transformations:

> When Charlie was six months old, he was inadvertently shut inside a refrigerator for five hours. Somehow, that event seems to have been a turning point in his life — transforming him virtually overnight into a prolific painter.

As soon as Minnie left Lyon and went to live at the little vineyard in Aix-en-Provence, her paintings changed dramatically, and so did the reviews.[114]

This book mocks the very idea of animal art by grossly overstating the case for it. It has a make-believe bibliography to show how seriously the authors studied their topic, with titles such as *Paws for Thought: The Magic & Meaning of Litter Tray Relief Patterns* and *Why Dogs Don't Paint*. The final chapter analyzes destruction of upholstery as a form of artistic expression.

There are, however, serious studies of intentional visual art by animals. Some of these are being conducted in the field, such as the observation of bowerbirds mentioned earlier. Others have used a rather anthropocentric approach by handing our closest relatives the tools of the painter.

Apes with an Oeuvre

First there was the ancient Roman myth of Dibutade, who did the next best thing to taking a Polaroid: before her lover left on a long journey, she recorded his face by tracing his profile on the wall. But in 1942, in a letter to *Nature*, Julian Huxley gave us the contemporary origins-of-art story. He had observed a gorilla at the London zoo carefully track the outline of his own shadow on the wall. The gorilla did so thrice, and Huxley recognized "a relationship to the possible origins of human graphic art."[115]

Nadie Kohts watches her
young chimpanzee, Yoni,
draw with pencil on paper, in
Moscow, 1913. (Reproduced
with permission of Oxford
University Press).

It is only logical that the quest for the origin of the artistic impulse brought us to the ape. Others had made similar observations before Huxley. In the 1920s, in Moscow, Nadie Ladygina-Kohts studied the perception of shape and color in her young chimpanzee, Yoni, and watched him enthusiastically draw with pencils on paper. Experiments on ape art were also conducted in the 1940s at the Yerkes Laboratories by Paul Schiller, who pioneered a simple test: he marked pieces of paper with lines or shapes and gave them to a chimpanzee, Alpha, to see what she would do with them. Alpha didn't simply splash paint randomly, but carefully heeded the markings, incorporating them in the end product. If Schiller put marks in three of a paper's corners, for example, Alpha would invariably scribble another mark in the fourth.[116] Yet the full extent of ape art became known only after an artist/ethologist began to pay attention, in the 1950s.

Desmond Morris, author of the all-time popular-science best-seller *The Naked Ape* (1967), as well as of many other works, has been a pioneer of the burgeoning genre of literature—of which the present book is an example—that compares human and animal behavior. He has proposed many provocative ideas, for instance, that human talk serves the same function as primate grooming, and that the invention of marriage was a necessary step when our forebears began to hunt in groups because it helped regulate male competition. Perhaps because some scientists consider him a mere vulgarizer, they have elevated some of his ideas to theory without so much as a nod to the author who inspired them. Morris started his career as a serious and respected ethologist, however, training at Oxford University under Niko Tinbergen.

Morris is also a surrealist painter in a style reminiscent of Miró. Art may even be his first love, and his paintings have been featured in several illustrated books and at major exhibitions. It was his sensitivity to art, combined with his opportunity as a zoologist to interact with Congo, a young chimpanzee, that provided Morris with rare insights into the nature of the artistic impulse.

Congo became a regular guest on Morris's television show, *Zootime,* and reached fame with an exhibition of his work in

Complex fan-pattern painting by Congo, a chimpanzee widely recognized for his excellent taste in color and sense of balance. Like other painting apes, Congo showed great concentration on the job, and was visibly annoyed if anyone tried to remove his work before he was done. (Photographs by Desmond Morris, both reproduced with his kind permission).

1957. His paintings were not merely a curiosity: they were widely recognized as beautiful. Congo had a refreshingly energetic style, and he seemed to strive for symmetrical coverage, rhythmical variations, and eye-catching color contrasts.

Congo stayed within the borders of the paper, never going off the edge, and made rudimentary compositions, such as a heavy dot surrounded by bold circular strokes, or a fan-shaped widening of lines. His art was considered beyond the level of that of a young child in terms of both composition and artistic boldness. The latter may have been due to the fact that chimpanzees are physically stronger and have better motor control than young children. Their paintings immediately strike us as forceful statements, whereas a young child's art tends to look tentative and hesitant.

Picasso hung a Congo on his wall. The paintings of other apes—one of whom was named Pierre Brassou to trick art critics—have been accorded serious, sometimes glowing reviews by experts who, unlike Picasso, thought that the artists were human.

One illustration of the power of ape art is how hard it is to emulate. Thierry Lenain, a Belgian art philosopher, recounts in *Monkey Painting* how an Austrian painter, Arnulf Rainer, tried to copy each and every body move and brush stroke of a painting chimpanzee. In 1979, Rainer squatted next to the ape, hoping to produce works of the same clarity and intensity. The human painter, however, evidently had the preconceived notion that apes are wild creatures devoid of emotional control. As a result, instead of imitating the ape, Rainer acted the way he *thought* an ape would paint. But he had it all wrong; apes can be as concentrated and controlled as people. As Lenain's account of a filmed session shows, it was the human painter who got too wild for the ape's taste:

> We see [Rainer] in the grip of a kind of trance, banging the paper, spitting on it, waving his brush nervously, throwing it down. The chimpanzee by contrast paints peacefully to start with, but is gradually influenced by the agitation of its imitator. It stops drawing, starts jumping about energetically and chases Rainer across the room. . . . Painting is not a violent activity for chimpanzees.[117]

If the ape's owner had not put an end to the pursuit of Rainer, the painter might have learned that an ape, even a young and relatively small one, has the muscular strength of several grown men bundled into one. Hence, an ape can charge a painting with energy and rhythm with far less effort than a person can.

In addition, ape painters don't seem to follow the rules that human artists do. Instead of worrying about the cumulative impact of an entire series of brush strokes and dabs, apes give the impression of taking a kinesthetic and visual pleasure in each separate action. We don't know the aesthetic secrets of the chimpanzee that Rainer tried to imitate, but the fact is that the human painter failed miserably in his attempt to achieve the same directness and sovereignty of expression. When Lenain examined fifteen works simultaneously produced by ape and human, he concluded that "[t]he chimpanzee's compositions are straightforward and clear. The imitations, on the other hand, are fuzzy, tangled webs of lines, completely illegible, almost to the point of hysteria."

The title of the English translation of Lenain's book, *Monkey Painting*, is unfortunate because, apart from a capuchin monkey named Pablo, all major nonhuman primate artists have been apes.[118] But the book contains an intriguing theory of primate art that is dramatically different from the ideas of Desmond Morris, who emphasized the similarities between ape and human. Lenain stresses the differences, and looks at ape art as a form of visual disruption. He believes that the painting ape disrupts the empty white space in front of him or her, testing and probing, and ultimately destroying what existed before. In contrast, Morris recognized a sense of aesthetic order and balance in the works of apes.[119]

Morris's art-as-order hypothesis has major points in its favor. First of all, apes seek a balanced and orderly arrangement in their paintings. Following Schiller's lead, Morris would place a mark off center, say to the left, and give the paper to Congo. Congo would tend to balance the composition by painting on the right side of the paper. He was not simply attracted to the empty space there, because the closer Morris placed his mark to the center of the page, the closer to the center on the other side Congo painted; and the farther to the left Morris put the mark, the farther to the right Congo worked, to keep the painting balanced.

Another indication that apes do not just make disruptive marks comes from the fact that they have a sense of completion of a painting. This is in contrast to what some early observers claimed. They argued that ape paintings are actually a human product: apes happily paint away until the product starts to look like a piece of abstract art to the people around it, who then take it away from the ape and hang it in a gallery. That would mean that the art is all in the human eye, that an ape has no conception of making a finished product.

But coming between an ape and his or her work can be dangerous! There are many stories of apes vehemently objecting to interruptions before they have finished their paintings. For example, Bella, a chimpanzee at the Amsterdam Zoo, painted with great concentration and was generally extremely peaceful until she lost her temper one day—with dire consequences for the keeper who tried to remove her materials in the midst of artistic activity. Morris also reported that Congo became greatly annoyed if he saw that a painting on which he was still working was about to be removed; nor did Congo like to be urged to continue once he had put down his brush, indicating that he was done. One day, Morris managed to take away a painting of an incomplete fan shape. When Congo got it back a while later, he simply continued where he had left off, carefully finishing the pattern.

A telling experience is that of Lucien Tessarolo, a French painter, who used to work side by side with a female chimpanzee, Kunda, on a canvas that both of them would sign at the end—Tessarolo with a signature, Kunda with a handprint. Tessarolo was impressed by Kunda's precision and harmonious choice of colors. The figurative elements that he added to their work were not always appreciated by the ape, however. Sometimes she reacted enthusiastically, but on occasion she rubbed Tessarolo's contributions out and waited to continue painting until he had come up with something else.

That doesn't sound like an ape seeking to disrupt order. Underlying Kunda's behavior must have been a sense of how the completed product should look. I am not saying that the product represents much value to the ape once it has been brought about, or that destructive tendencies never occur. Indeed, as soon as the production phase is over, apes have been known to tear their works to shreds. On other occasions, they have exhibited an indifference to their finished paintings that humans find hard to understand. In this regard, the apes are very different from human artists: their goal is not to create an enduring visual image that will please, inspire, provoke, shock, or produce whatever effect it is that the human painter seeks to achieve.

The evidence, then, is that painting apes have a sense of both balance and completeness, enjoy the visual effect of what they do, and create regularities and patterns, but are not out to produce a lasting product. As far as they are concerned the product can be thrown away once they are done with it. So, even though their painting activity is best described as the deliberate creation of visually pleasing patterns, rather than as a form of disruption, it differs from our artistic activity in that it does not appear to be a means to an end.[120]

The Germ of Aesthetics

Some people feel that calling what apes produce "art" mocks human achievements. Indeed, the use of primates as caricatures of ourselves has a long history, including an entire genre of seventeenth- and eighteenth-century art depicting capuchins or macaques sitting behind an easel, paintbrush at the ready, staring at a female nude or still life just as a human painter would. Whether those paintings were a commentary on the slavish copying by some human artists, or self-mockery by the painters, the underlying message was one of opposition between animal and art. If art is by definition a human domain, a monkey with a paintbrush can only be a joke.

The age-old "monkey artist" theme had to pop up, of course, when ape paintings became an issue in the 1950s. One famous chimpanzee, Baltimore Betsy, was customarily photographed in front of her work with captions such as "Just a little something I dashed off, but not bad." Such catering to the general public's sense of humor undermined any attempt to explain what is interesting about ape paintings.

Moreover, apes came in handy in a cultural war zone of the time concerning gestural art and action painting characterized by vigorous, dynamic brush strokes and random effects of spilling and dripping. Since ape art looks similar, it became a weapon against these schools, with critics expostulating that if an ape can do what certain human artists are doing, the humans must be operating at a rather primitive level. Salvador Dali, for example, couldn't resist making the following calculated jab at another painter: "The hand of the chimpanzee is quasi-human, the hand of Jackson Pollock is almost animal."[121]

People accused Morris of trying to ridicule modern art, but that was never his goal. If people get over their giggles and consider the issue at hand, they will see that there is a serious question behind it. Why do the members of our species all over the world produce art? What is it that drives them? Why waste time and energy on this sort of activity? Is it a form of play, a form of exploration, a mental game, a way of impressing others? Morris simply wanted to show that we are not the only species to take pleasure in self-created visual effects, hence that the aesthetic sense probably has older roots than is often assumed.

But where does ape art end and human art begin? The main dividing line seems to be representation. In spite of isolated claims of apes producing recognizable images (for example, Koko, the gorilla, is said to have painted a bird, a dog, and a toy dinosaur), I have never been able to recognize the purported images in their paintings. Human art seems to me unique in its depiction of reality. Observing that the human child moves on to representations after an abstract phase, Morris and his wife, Ramona, concluded in *Men and Apes* that "unhappily this is the point at which the apes get stuck."

Yet even if substantial differences between human and ape art remain, they should not distract us from the undeniable common ground. Obviously, we feel that there is more to our art than enjoyment of visual effects: the human artist imagines and strives for an end product. Human art is a conscious act of creation. On the other hand, without satisfaction derived from intermediate stages—from the activity itself and its immediate results—we might never have reached this point. It is in this regard that ape art, rather than insulting our ego, provides a glimpse of the wellspring of the universal human artistic impulse.

What Is Culture

and Does It Exist in Nature?

If culture is the transmission of habits and information by social means, it is widespread in nature. Animals may have no language or symbols; but they develop new technologies, food preferences, communication gestures, and other habits that the young learn from the old (or the other way around). As a result, one group may behave quite differently from another, and culture can no longer be claimed as an exclusively human domain.

Despite abundant evidence for this idea, there exists enormous opposition to it. Counterclaims focus on the learning process, which most of the time seems rather simple compared to human cultural transmission, or on the peak achievements of human civilization, with the remark that nothing of the kind is within the ape's reach. Animal culture is further-

more denied the survival value of human culture. Research carried out over the last few decades suggests, however, that the survival of many animals in the wild hinges on what they learn from others. They take advantage of accumulated knowledge, and in this sense are as reliant on culture as we are. Underlying the learning process is the same desire to belong and fit in that we recognize in our own cultures. And as with human culture, the variations brought about are not indefinite; they are built around the shared heritage of the species.

5

Predicting Mount Fuji

and a Visit to Koshima, Where the Monkeys Salt Their Potatoes

"A scientist is a man who by his observations and experiments, by the literature he reads and even by the company he keeps, is pulling himself in the way of winning a prize; he has made himself discovery prone."

Peter Medawar, 1984

"Instinct is an inherited behavior and thus is something opposite to culture, which represents acquired behavior. If it is dogmatic to regard all animal behavior as instinctive, it is equally dogmatic to regard all human behavior as cultural."

Kinji Imanishi, 1952[122]

A Japanese scientist hikes around central Honshu and suddenly, high up in the sky, sees something that looks like a shimmering mountaintop. If it is indeed a mountain, most of it remains invisible, and an hour later the top has vanished behind the clouds. Can a mountain walk away?

The next day, he sees the entire Mount Fuji. With its 3,776 meters, it is Japan's highest peak, dwarfing everything around it. The scientist devotes his life to describing the mountain and its unparalleled steepness and perfectly symmetrical cone shape, photographing it from different angles in a fashion similar to Katsushika Hokusai's famous nineteenth-century print series *Thirty-six Sites in Edo Overlooking Mount Fuji*. The sacred volcano is there to be admired and worshipped. Surrounded by temples and shrines, it is a place of pilgrimage for millions of Japanese.

For the Western scientist the mountain comes equally unexpected, yet he finds this a bit disturbing. Shouldn't he have known that this thing *could* be here? The scientist proceeds to study plate tectonics, volcanic eruptions, magma viscosity, and so on. After much reading and exploration, Mount Fuji appears logical, almost inevitable. The scientist is satisfied that this mountain's dazzling appearance high above the landscape is not nearly as surprising as it seemed at first. Instead of offering different views of the mountain, he publishes a thorough treatise about the origins of volcanoes. His theories have led him to believe that eruptions are predictable, and to prove his point he explains how he set foot on Honshu expecting a large volcano somewhere in the vicinity of Tokyo. Lo and behold, about 100 kilometers west of the city he found the predicted mountain. This verification is a testimony to his superior science.

But isn't this scientist cheating? Didn't he see the mountain *first*? Yes, but this is a customary way of presenting evidence. Facts somehow look better if anticipated, and scientists sometimes achieve this result by formulating rules that fit the facts that helped them develop the rules.

Such is the baffling, circular state of affairs in the behavioral sciences. For example, an American primatologist, Jeffrey Kurland, once set out to test kin selection theory, which predicts friendly and supportive relations among close relatives.[123] His 1977 monograph on the behavior of Japanese macaques opens with purely theoretical considerations, such as evolution by natural selection, the transmission of DNA, and the sharing of genes between relatives. Equipped with a set of logically compelling predictions, Kurland seems to be entering a new area of investigation. Would he prove the theory right? The reader can't wait to hear the results, sure that they will resolve many burning issues.

The only flaw in this presentation is that the study required a troop of monkeys with known genealogical relations, which means that all individuals were known and that careful records had been kept, year after year, about who gave birth to whom. But why in the world would anyone ignorant of the role of kinship have taken such trouble? Kurland found his troop at Ryozenyama, in the Suzuka mountains, where Japanese scientists had such records going back to the 1950s. They had started this project well before William Hamilton's classic paper on kin selection, which appeared in 1964, knowing that kinship in macaques dictates social rank and bonding. Kurland did an excellent job documenting the close relations among kin, and in doing so produced a more detailed picture than existed before, but he could only conduct his study because what he set out to demonstrate was already largely known.

The urge of behavioral scientists to proceed in a straight line from theory to data, hence presenting themselves as more naïve to the truth than they actually are, derives from a desire to be like physics, a science that has reached the lofty stage of armchair prediction. Soon after a new theoretical insight has been achieved—that there must be quarks, or that the collision between a meson and a proton should produce a lambda—hordes of scientists set out to verify its predictions in the enormous accelerators and bubble chambers of CERN, near Geneva, or Fermilab, near Chicago.

It is unclear whether the behavioral sciences will ever reach the point when logically derived predictions drive progress. Behavior is more variable than the dance of photons, and its explanation involves multiple layers, from the physiological to the mental. We cannot afford to look through a single pair of glasses; we need lots of different glasses to see reality. Theories do assist in this effort, by guiding our attention and making large amounts of data graspable, but they also induce selective blindness.

Theories are often formalized, but there is no reason to deny the relevance of general expectations. Without openness to the idea of animal culture, for example, the potato washing by monkeys on Koshima Island might never have attracted any attention. What's the big deal of monkeys running to the ocean to dip spuds into it? Once the question of animal culture was formulated by Japanese primatologists, however, this simple behavior took on enormous significance, inspiring them to take careful notes for decades.

The question at hand, then, is what people see, or do not see, as a result of preconceived notions. Instead of predicting new events with great precision, or explaining what has already been found, a great deal of scientific progress is linked to what is deemed possible and likely. Expectations, however vague and intuitive, lay the groundwork for discovery: every new insight slowly grows under the surface of human consciousness before it bursts into the open. If no one had formulated a broad concept of culture, it is unlikely anyone would have been looking for culture in animals.

Respect for the Unexpected

When people still believed that the earth was flat, they overlooked or explained away signs, such as the presence of a horizon, that might have told them otherwise. The unexpected often escapes attention. Similarly, lots of young animals were killed by their own kind, on farms and under the eyes of naturalists, before someone dared to call infanticide a pattern. The behavior fills us with horror, and it flies in the face of the idea that organisms do their utmost to survive and reproduce. Killing of new life didn't make sense, and consequently didn't register.

In reality, infanticide is not that unusual. In 1967, Japanese primatologist Yukimaru Sugiyama reported how male langur monkeys in India, when taking over a harem of females by ousting the old leader, customarily kill all infants in the troop. They snatch them from their mothers' bellies, mauling them with their sharp canines.[124] The very first presentation of these findings for an international audience met with a deafening silence, followed by dubious praise from the chairman that "Dr. Sugiyama has offered us some intriguing examples of behavioral pathology." But the investigator himself never spoke of pathology, and to this day says he has no idea what it means. Animals respond in various ways to various conditions; it doesn't help in any way to call one way normal and another abnormal.

The discovery was ignored for about a decade, after which other reports of infanticide surfaced, first in other primates and eventually in many other animals—from lions and prairie dogs to dolphins and birds. I have never witnessed such turmoil at primatological conferences as in the days when infanticide became a growing topic. Reports provoked shouting matches, accusations of inadequate evidence (most of it was postmortem), and utter disbelief that the same theories that speak of reproductive success could be enlisted to account for the annihilation of newborns.

But this is exactly what happened, beginning with Sugiyama's own suggestion that the loss of an infant induces a female to be ready to mate sooner than otherwise possible, and that this might be good for the male. Thus, not only did the investigator report a disgusting phenomenon, but he had the temerity to suggest that it might exist for a reason. This idea was followed by formulations that stressed how a male, by removing the progeny of another male and mating with the female, may increase his own reproductive output. If so, the infanticidal tendency will be passed on to his sons. Based on almost twenty years of data from wild langurs, there is now indeed excellent support for the idea that infant killing represents a male reproductive strategy.[125] Infanticide is increasingly regarded as a key factor in social evolution, pitting male against male and male against female. Females have nothing to gain: the loss of an infant is a tremendous waste of maternal investment in gestation and lactation.[126]

Given that the unexpected is inherently more exciting than the expected, the high status of theory testing remains a bit of a puzzle. The discovery of infanticide, and our slow realization of its implications, goes to show how the greatest advances in science occur when precooked ideas fall short, forcing us to come up with a fresh perspective. As explained by Ernst Mayr: "In biology, concepts play a far greater role in theory formation than do laws. Two major contributors to a new theory in the life sciences are the discovery of new facts and the development of new concepts."[127]

If perceptivity and curiosity are indeed critical for scientific progress, why don't we in the behavioral sciences teach our students to keep their eyes peeled? Instead, we urge them to develop hypotheses, list them faithfully in the introductions to their theses and papers, then devote the rest of their work to a demonstration of the correctness or incorrectness of their predictions. Not surprisingly, students develop a knack of writing introductions that make them look prescient: it is often hard to tell whether their predictions fit the facts or the other way around.

There is really not much against such neat presentations so long as we realize that they are only that—a way of organizing and presenting our work. The structure of papers is not to be confused with the actual process of science, with its detours, surprises, and frustrations. Lewis Wolpert, in *The Unnatural Nature of Science*, calls the scientific paper "a kind of fraud," but I rather see it as a collective lie: we all know how things really work.[128]

The student who sets out to test a single idea is bound to return either disappointed—the behavior may not occur often enough, or the idea may prove naïve or wrong—or without having learned much due to an excessively narrow focus. For example, an aspiring field-worker may follow elephants for years to prove that cows mate with the largest bull, but if all she learns is related to who mates with whom, she will never amount to much of a scientist. A much broader orientation is expected, an appreciation of the elephant's ecology and social organization, leading to new observations and new ideas inspired by what these animals actually do rather than what they are supposed to do.

Most scientists know the fertile interchange between theory and observation: the first inspires the second as often as the second the first. Charles Darwin is a case in point. He developed his grand theory only after having sailed halfway around the globe, collecting information, sitting on it, rearranging it in his head, until all of it seemed to fall into place. Darwin was a naturalist long before he turned theorist.

What a contrast with a young scientist I once encountered who had decided to become a theoretician. He had never done any actual research on either animals or people, but told me with astonishing aplomb that his ambition was to develop theories that would revolutionize the behavioral sciences. "I want to give other scientists something to work with," he said, evidently viewing those of us who do the nitty-gritty of data collection much as the queen bee views her workers. Every day, he sat behind his computer, never leaving his desk, a superhuman effort comparable to a novelist who has never set foot outside the house, never lived a life, but aspires to write as richly as Tolstoy or Dostoyevsky.

I am still waiting for his egg to hatch.

Noble Savages

Having explained that Western approaches to animal and human behavior are not nearly as theory-driven as advertised, it is interesting to look eastward. By this I do not mean all of Eastern science, but just the little corner that I know best. Off the record, Western colleagues used to complain about the "lack of theory" in Japanese primatology. The emphasis was on data gathering—for example, about what monkeys eat or whom they groom—without mention of the idea behind it. Data without a framework to put them in seemed pointless.

Looking beyond the surface, however, there can be little doubt that Japanese primatologists had plenty of assumptions. They may not have advanced these as formal theories, but their observations were never made in a vacuum. And what is more, their assumptions eventually triumphed. They are now shared to such a degree by everyone else that we consider them conventional wisdom!

Until well into the 1960s, chimpanzees were seen as Rousseau's noble savages: they traveled autonomously and self-sufficiently around the forest in haphazard combinations. The ever-changing parties gave the impression that, except for the mother-child relation, these primates lacked any enduring ties. Thus, Jane Goodall called females and their dependent offspring the only stable social units. Working only 130 kilometers south from her site, also in Tanzania, a Japanese team under Jun'ichiro Itani and Toshisada Nishida had the same trouble getting a good overview of chimpanzee society. From the outset, however, they assumed that they were dealing with highly social beings. Familiar with the tightly organized macaque troops of their native country and guided by a cultural emphasis on collectivity rather than individuality, they recognized the survival value of group life, the role of social transmission, and the need for each individual to belong. They believed in social connectedness. How could a species that is supposed to fill the gap between the monkey and us have no society to speak of?

Eventually, through persistent observation, they cracked the puzzle and showed that chimpanzees live in large communities, the membership of which is stable, especially for males. As opposed to many other primate species, in which males migrate between groups, they discovered it was the females. The male-philopatric society of the chimpanzee is now taken for granted—we all have heard about territorial wars between different communities and about group-specific traditions—but the initial discovery came out of a firm conviction that chimpanzees could not be nearly as individualistic as Western science had made them out to be.

In a revealing passage, Nishida describes how he greeted his mentor, Itani, when the latter came to visit Africa in 1966. There had been much debate among Japanese scientists about the social life of gorillas, which live in harems of one or two males with a couple of females. Could these family units, which they dubbed "familoids," be the ancestral type of human society? If so, chimpanzee society should follow the same model. Nishida, who stood onboard a steamship, couldn't wait to give Itani, who was waiting on land, the verdict from the field. He shouted: "No familoid in a large-sized group!" Itani countered, "It cannot possibly be the case!"[129] This is hardly the sort of exchange expected between people without a theory.

The Individual in Society

Plato's Great Chain of Being, which places humans above all other animals, is alien to Eastern philosophy, according to which reincarnation of the human soul can occur in many shapes and forms. This means that all living things are spiritually connected. A man can become a fish, and a fish can become God.

The presence of monkeys in India, China, and Japan—in contrast to the Middle East and Europe—may have strengthened people's closeness to nature: seeing other primates makes it hard for us to deny that we are part of nature. In the same way that Western fairy tales feature animals such as foxes, rabbits, and ravens, Eastern folk tales and poetry are laced with references to the monkey as a mirror of ourselves. Monkeys are held in the highest esteem. The three wise men, or magi, of the bible are matched in the East by the three wise macaques of Tendai Buddhism ("See No Evil, Hear No Evil, Speak No Evil"). The Japanese language has a special honorific affix for the monkey: *o-saru-san*, or "mister monkey."

Humility in the face of our close relatives has obvious implications for the way we study them. Without a religion that
grants souls to only one species, neither anthropomorphism
nor evolution stirs up controversy. If the soul can move from
monkey to human, and back, there is no ground to resist the
idea that our species are historically linked. On the contrary,
evolution seems a logical and welcome thought.[130]

Given this background, Japanese primatology had from the
start its own unique set of attitudes and concepts. For one
thing, animals were considered as individually different as
people. Many great advances can be traced to the habit of giving each individual a name. And not only were the individuals
told apart by external markers such as face, size, and color, but
there was the clear implication that they had personalities just
like people, and that each should be understood on its own
terms. Needless to say, the naming habit was frowned upon by
Western scientists, who felt that it unduly humanized animals.

All of this happened in the early 1950s, at a time when
European ethologists were interested in instincts and species-
typical behavior. For this purpose, one doesn't need to recognize individuals; it is sufficient to watch how animals of a particular species respond to particular situations. American
behaviorists, on the other hand, stressed general laws of learning that were supposed to hold for all animals, including humans: not even the species was a significant entity, let alone
the individual. One can see the enormous contrast with
Eastern scientists, for whom everything revolved around the

society and the place of the individual in it. They set themselves the task of documenting each individual's kinship relations, friendships, rivalries, and rank position. *Connectedness*
was the key: connectedness among all living things, including
ourselves, and connectedness among all members of society.

Unity with nature led to another staple of the Japanese approach: food provisioning. Shintoistic nature worship expresses itself in the feeding of animals, both tame and wild,
such as carp, birds, deer, and monkeys. Shinto priests leave of-
ferings of fruits, rice, and drinks at their shrines for the *kami*
(spirits of the natural world); monkeys are believed to come as
intermediaries to pick up the food.

The food provisioning adopted by scientists as a way of getting close to their study objects thus derives from an age-old
tradition. For Japanese scientists, provisioning served their research, but it also created a bond. As explained by Pamela
Asquith, a Canadian anthropologist specializing in the comparison of Western and Japanese primatology:

> Historically, there was another aspect to provisioning the
> macaques for some Japanese researchers. Through feed-
> ing, a relationship of natural empathy was created be-
> tween observer and monkeys. Masao Kawai considered it
> another part of their methodology, and termed it the
> *kyōkan* (sympathetic or responsive) method. By feeding,
> the researcher positively entered the group and made con-
> tact with the monkeys. At the time when most Western re-

searchers advocated strict neutrality with study animals, this was indeed unique. . . . Artificial feeding in Japan was never done with the sole purpose of "making friends" with the monkeys, but for some, it lent another, psychological, dimension to provisioning.[131]

In 1958, Kinji Imanishi, whom I discussed in Chapter 2 as the father of Japanese primatology and defender of a harmonious world view, toured American universities together with some of his students to report the first findings from their detailed research. They encountered much admiration but also a great deal of skepticism about their ability to distinguish all those monkeys and track them over time. The greatest American primatologist of those days, Ray Carpenter, became a staunch supporter of Japanese primatology, however, visiting Japan three times in the years to come. Whereas he himself had assigned a much less prominent place to individuality and personality in his work, the advantages of such an approach were immediately obvious to him.

Within a decade, both individual identification and food provisioning were adopted at Western field sites, from Gombe Stream to Cayo Santiago. And even though the second part of the method has fallen out of favor, the first remains firmly embedded. The concept of the individual in its society amounts to a momentous theoretical contribution by Japanese primatology to the study of social animals. The ideas that individuals matter, that their identities are linked to their place in the whole, that they need to be followed over time, and that human empathy helps us understand them, are so obviously correct that armies of scientists now apply this perspective, often without knowing where it came from.

The Prepared Mind

I saw Fuji-San for the last time on my way from Hokkaido, Japan's northernmost island, to Kyushu, in the south. I boarded the plane in Sapporo, where the arrival of a long and cold winter could be felt. The climate, landscape, and flora (not to mention the so-called "Bier Gartens") reminded me of Germany. I flew a distance equivalent to that from Hamburg to Rome toward a balmy climate, palm trees, and spicier food. Somewhere midway, the lonely, snow-covered top of Mount Fuji appeared on the left of our plane while its base remained under the cloud layer—a rare sight, and a magical birthday treat. Without friends and family to remind me of my aging process, I was sort of hiding from the big Five-O.

I was on my way to Koshima, the tiny island off the coast of southern Kyushu where monkey research began around the day of my birth. A couple of years later, in 1953, the cultural revolution of primatology was set off by a major discovery on the island: the spontaneous occurrence of potato washing by macaques. Here again, observation was preceded by expectation. In a book that criticized the Western view of animals as automatons driven by instinct, Imanishi had inserted an enlightening discussion between a wasp, a monkey, an evolutionist, and a layman, in which the possibility was raised that animals other than ourselves might have cultural transmission.[132]

This is really all that is needed: a certain preparedness to interpret reality one way or another, and to recognize which findings fit or don't fit prevailing views. Free from hang-ups about human uniqueness and the primacy of the individual, Japanese primatologists were mentally primed for a simple observation that forever changed our field.

Koshima

Macaques have a special call to announce a welcome change in circumstances—when the rain stops after a storm, for instance, or when, on a cold day at the zoo, they hear the door to their indoor quarters being unlocked. They give this happy "coo" call also when they anticipate food, but I had never heard it so loudly.

Walking to our boat on the pier, on the way to Koshima island,[133] monkey calls reached us over the noise of the Pacific. The shrill sounds came from above, from lookout places on the tall, forested rock of Koshima. The island is so close to the coast that we were within sight. The monkeys must have recognized the field assistants and noticed the two heavy bags they were carrying. They are also said to recognize by ear the research boat's engine. Given their noisy advertisement, it wasn't surprising that most of the one hundred monkeys on the island had gathered when we landed on the beach.

Compared to the "snow monkeys" of the north, the monkeys here had thinner and darker coats. They were also smaller than any Japanese monkeys I had seen. It is estimated that this island can support only around thirty monkeys with natural foods, such as acorns and leaf buds. Due to human provisioning, since the 1950s the population had kept growing and growing until the researchers, in 1972, decided to seriously curb feeding. Small handouts of wheat are still being given several times per week, but sweet potatoes are now fed less than five times per year.

I was lucky to be present for one of these rare potato feedings. Potato washing on Koshima has reached textbook status as an example of animal culture, but in some circles its implications have become suspect. I needed to see it with my own eyes.

It is strange to think that research on Japan's native monkeys, which started at this very spot, was only a second choice, a felicitous outcome of boredom with hoofed mammals. The history goes back to Cape Toi, a peninsula of southeastern Kyushu, where Kinji Imanishi and his students from Kyoto University were observing "wild" horses. I put "wild" between quotation marks because when I went to Cape Toi, the horses let themselves be patted on their heads. Before the Japanese had the means to travel to Hawaii or Australia, this subtropical place with its beautiful view of the mountainous coast was *the* honeymoon destination. The horses are completely used to tourists, who undoubtedly represent a great improvement over the heavily harnessed samurai whom their ancestors had to carry into battle.

According to Imanishi's most prominent student, Jun'ichiro Itani, the horses of Cape Toi were totally uninspiring: about all they did was graze, sleep, and move about. At the same time, another student, Shunzo Kawamura, studied the famous deer of Nara. Nara is the old capital of Japan, from before Kyoto, and well before the present capital. Having roamed Nara's temple grounds for over a thousand years, mingling with visitors as they please, the deer, too, are terribly tame. I experienced this firsthand in an embarrassing incident resulting from a tempting childhood image.

The Todai-ji temple in Nara is an enormous wooden building devoted to the world's largest bronze statue, known as the Daibutsu, or Great Buddha. During a visit, I noticed a stand near the temple that sold stacks of flat, round cookies exactly the same color, size, and shape of Dutch *stroopwafels* (syrup wafers), which are sweet and delicious. For all I knew, given the ancient Dutch influence on Japan, they might have derived from the real ones. I couldn't resist buying these cookies, but before I had even handed money to the woman behind the stand, a deer found me among the crowd and pulled with her teeth at my shirt. Another gently poked me with his antlers in the back. Both clearly wanted the pastries, and they were so insistent that I had to run holding my prize aloft. After escaping this way—to the amusement of onlookers, no doubt—I finally managed to take a bite. My taste buds immediately sent me the news: these cookies were *not* for human consumption! Apparently, every soul in Japan knows that they are for the deer, as do the animals themselves.

Kawamura's deer weren't thrilling study objects either. Hence the little wordplay of Itani, who likes to point out that since the Japanese word for horse is *ba*, and for deer *ka*, scientists were bound to find out one day that to study these animals was *baka-rashi*, which means "foolish." This profound insight reached Itani and Kawamura when, while taking a break from horse watching at Cape Toi, they saw a procession of monkeys far away in the sunset. They were mightily impressed by the apparent organization of the moving troop and by the monkey's exchange of calls, which indicated a refined communication system. In the evening, in their old inn, the students discussed with their master how it would be to work with monkeys. Wouldn't it be a lot more fun than working with horses? Their excitement won, and the decision was made to visit Koshima. Already at the time, this island had the status of National Treasure, which meant that the monkeys there were protected. There was also less tourist disturbance.

The next day, they began their journey. Nowadays, with cars and highways, it is only a short distance to Koshima, but it took them a full day by foot. To get to the island, they needed a boat that would be available the next day. They stayed overnight with a farmer who had strapped the arm of a monkey over his horse stable to ward off evil spirits. Following this pointer, they traveled on to the island, which they crisscrossed in all directions—getting lost and all—without seeing any primates except for each other.

But they did notice monkey droppings.

Imo's Innovations

This first exploration took place on December 5, 1948. Soon thereafter, provisioning with wheat and sweet potatoes began so as to habituate the extremely shy monkeys to people. This technique, which has the same Shintoist roots as the feeding of Nara's sacred deer, allowed the daughter of the farmer with whom they had stayed, Satsue Mito, to begin the arduous task of identifying individual monkeys, giving them names, and describing their social network.

Half a century later, Mrs. Mito is nationally famous as the author of books about her work. During my stay, in the fall of 1998, I discussed monkey matters with the eighty-four-year-old through an interpreter in her own *minshuku,* a traditional simple inn. Mito's *minshuku* is in the village of Ichiki, close to the ocean. It is popular for its bath house, which draws water from a natural hot spring. Over delicious homemade dinners, she talked with me about her monkeys with great love, like a grandmother about her grandchildren, remembering every face and name, a glint in her eye for some, a sad look for others.[134]

There was the time when a female, Aome, fell in love with a young researcher. She would literally cling to his leg when he moved around the island. Later another female, Imo, had the same attraction, and would sit on his shoulder. When this beloved researcher visited the island after an absence of six years, Imo remembered him and was particularly nasty toward his wife. Or, there was Utsubo, who, after her infant died, carried the corpse around until it began to decay. Such attachment is not unusual in mammalian mothers, except that Utsubo kept the desiccated body for no less than fifty-nine days! Then there was the occasion, in the postwar years, on which an American army commander wanted a pet monkey and asked villagers to hunt on the island. We also looked at pictures of twin monkeys born on the island, of a researcher who later drowned in a typhoon, of a visit by the American primatologist, Ray Carpenter, and of the slow deterioration of a female monkey with breast cancer.

Finally, there was the occasion, in September of 1953, on which Mito noticed how Imo, then an eighteen-month old juvenile, carried a sweet potato to a small freshwater stream that ran from the forest to the beach. Eating soiled potatoes wears down teeth, so it seemed a good idea to clean them. Imo did so by rubbing the potato in the water. She playfully repeated this behavior on the first day. Later, she improved her technique by going deeper into the water, holding the potato in one hand and rubbing off the mud with the other, occasionally dipping it in the water. No pictures remain of these events both because they happened unexpectedly and because their significance was fully realized only later on.

Mito didn't wait long to send a letter to distant Kyoto, where it was read by Imanishi and his students, including Kawamura and Masao Kawai. Very soon, Kawamura began collecting information resulting in a first Japanese article, in 1954.[135] But perhaps the best-known article on the Koshima monkeys appeared ten years later, in 1965. Kawai put the exalted C-word in its title, even though he softened the blow with a prefix: "Newly Acquired Pre-cultural Behavior of the Natural Troop of Japanese Monkeys on Koshima Islet."

The discovery of potato washing is therefore best looked at as a team effort. Mito noticed it first; the cultural interpretation came from Imanishi; and Kawamura and Kawai collected the necessary information to convince the world of social transmission. Only a year before Imo's giant leap for monkeykind, Imanishi had speculated in writing about the possibility of animal culture, succinctly defining the phenomenon as "socially transmitted adjustable behavior." Thus, Mito's letter fell on fertile ground: her observations must have clicked in Kyoto. It was the first sign that the standard view that animal habits are handed down exclusively by genetic means needed revision.

Kawai's report is early Japanese primatology at its best. It describes in admirable detail how potato washing first spread horizontally, from Imo to her playmates. Within three months, two of her peers as well as her mother were showing the same behavior. From these potato pioneers the habit spread to other juveniles, their older siblings, and their mothers. Within five years more than three quarters of the juveniles and young adults engaged in regular potato washing. Older males, however, failed to adopt the habit.[136] Kawai explains that transmission seemed to follow the amount of time monkeys spent together, and that because males over the age of four typically live at the periphery of the troop, they had little exposure to potato washing.

Transmission according to kinship lines, and from the young to the young and the young to the old (rather than from the old to the young), also occurred after a second innovation by Imo. In 1956, she introduced a solution to the problem that wheat thrown onto the beach mingles with sand. Imo learned to separate the two by carrying handfuls of the mixture to nearby water, and throwing it into it. Sand sinks faster than wheat, making for easy picking. This sluicing technique, too, was eventually adopted by most monkeys on the island.

Acquired Taste

All monkeys present during my visit had been born long after these events. Even the oldest individual, an alpha male named Noso, was "merely" thirty-one years old. That Noso was still in power was truly remarkable given that he looked positively worn out. In captive monkey groups, I have known similar situations of an old male clinging to the top rank. Under those conditions, potential challengers are usually from the same group. Since they have known the old boss since infancy, it is logical to expect a psychological inhibition to attack him. In wild macaques, in contrast, challengers enter the troop from the outside, and hence are totally unhampered by such respect. They make short work of an arthritic, feeble male like Noso.

But then, isn't Koshima a bit like a captive situation? All its monkeys spend their entire lives on the same island. In the annals of this place none of the alpha males has ever been overthrown: transfers of power have always taken place following the former alpha's natural death.[137]

Noso and the others eagerly followed us around until they understood that we were not going to feed them right away. In exploring the island, I stepped over the little stream in which Imo had dipped her first potato and was introduced to her descendants. In the forest, I encountered some of the young males who didn't dare come down to the small stretch of beach where the main troop gathered and relaxed. I also climbed the island's 113-meter peak: high by the standards of a Dutchman, but surely laughable to Imanishi, a mountaineer who organized expeditions to the Himalayas.

Leaning over an outcrop, I looked down on fishermen standing on rocks in the ocean, some with a monkey waiting by their side. In the best sushi tradition of this country, the Koshima monkeys have learned to eat raw fish, mostly by taking discarded fish from anglers. This was first observed in a few hungry older males, after which the habit spread to other adults, including females, and then to the rest of the population. The path of transmission was totally different, therefore, from that followed with both of Imo's innovations.[138] For all we know, these are the only Japanese monkeys to have acquired a taste for seafood. They also pry limpets from the rocks, and they have even been seen to capture octopi and fish trapped in small pools left behind during ebb.

Upon our return to the beach, the monkeys became quite excited. When the bags were finally opened, there was pandemonium and intense competition. While most of them chased each other around, quickly spreading the food to all corners of the beach, one male smartly sat down on a few potatoes while quietly nibbling on one. In the course of the hour or so that it took them to consume the food, I saw many monkeys run bipedally, both hands full, to the ocean. Walking in shallow water, they would alternate dipping a potato in it and chewing off a piece. They did not do much rubbing in the water, probably because these potatoes were prewashed: there was hardly any dirt to be removed. Soiled potatoes are not even commercially available anymore. For this reason, Japanese scientists have changed their terminology: they have stopped speaking of sweet-potato washing. Assuming that it is the salty taste of the water that the monkeys are after, they now speak of "seasoning."

Strange Rumors

In recent years, attention for these important discoveries has been diverted by two rumors. One sought to inject the events with a supernatural flavor, whereas the other derived from a skeptical reevaluation by a Western scientist.

In 1982, Ken Keyes published *The Hundredth Monkey*. Employing large lettering on almost empty pages and an abundance of exclamation marks, the booklet was designed to reach even the dimmest minds. Here are the contents of the critical four pages:

> Let us suppose that when the sun rose one morning there were 99 monkeys on Koshima Island who had learned to wash their sweet potatoes. Let's further suppose that later that morning, the hundredth monkey learned to wash potatoes.
>
> THEN IT HAPPENED!
>
> By that evening almost everyone in the tribe was washing sweet potatoes before eating them. The added energy of this hundredth monkey somehow created an ideological breakthrough!
>
> But notice. The most surprising thing observed by these scientists was that the habit of washing sweet potatoes then spontaneously jumped over the sea—Colonies of monkeys on other islands and the mainland troop of monkeys of Takasakiyama began washing their sweet potatoes!
>
> Thus, when a certain critical number achieves awareness, this new awareness may be communicated from mind to mind.[139]

The insertion "observed by these scientists" implies that the birdlike migration of the habit was well-documented. Keyes'

One of the strangest ideas about the spread of potato washing is that the habit could have jumped from Koshima to other islands after a critical mass of one hundred monkeys had learned it. (Cartoon by Rob Pudim, published with permission of The Skeptical Inquirer, in which it first appeared, in 1985).

source, however, was not one of the Japanese primatologists, but a New Age author, Lyall Watson, who chose to interpret the occurrence of potato washing at distant sites as evidence that habits can spread through thin air. But all of this was coincidence; there is no indication whatsoever that what happened at Koshima affected events elsewhere.[140] Keyes, however, had a political, not a scientific, agenda. Opposing nuclear weapons, he exploited Watson's misconceptions to argue for group consciousness. If enough people join the nuclear freeze movement, so his reasoning went, their collective mind-set will break down political barriers. Readers of *The Hundredth Monkey* were urged to learn more by contacting an organization aptly named "ClearMind Trainings."[141]

This leap-of-consciousness story is still doing the rounds in the American business community when motivational speakers want to impress audiences with the power of minds working together. Rumors tend to stay around, no matter how unsubstantiated. Since the same is true of criticism, I need to address here an influential scientific paper drenched in disbelief that monkeys might have culture. In 1990, Bennett Galef questioned whether the spreading of potato washing had anything to do with imitation.[142] The Canadian psychologist was right to take a close look at the evidence and to insist that scientists carefully weigh the options when they see a behavior spreading in a population. The learning process may range from simple to quite complex, and it is important to gather data that permit making the right distinctions.

But given Galef's valid warning, it was all the more disturbing that he himself made so little effort to verify his own assumptions, for example, by actually visiting the island in person. Instead of social learning, he suggested individual learning—that is, that each monkey had acquired potato washing on its own without the help or example of others. One pillar of this claim was an aside hidden in the primate literature according to which potatoes were selectively handed out by Mito to those monkeys who washed them. Is it possible that she had stimulated the monkeys to do what the scientists wanted to see?

Galef's source was another researcher, an American, who did visit the island. However, this visit occurred no less than fifteen years after Imo had introduced potato washing.[143] By that time, I was told, Mito would on occasion accommodate tourists and camera crews who wanted to document potato washing by feeding closer to the ocean and making sure that her best "performers" were at hand. All of this is irrelevant for the interpretation of how the habit started or spread: Kawai's analysis covered a much earlier period in which few outsiders ever showed up on Koshima.

But let us for a moment imagine that Mito did indeed reward potato-washing monkeys. How could this explain anything other than that monkeys who did it kept doing it? Reinforcement can strengthen an existing habit; it doesn't create one. Actually, withholding potatoes from those individuals who failed to wash them would be the best way to *prevent* potato washing from spreading. In other words, if Mito had done what Galef said she did, few monkeys would ever have had a chance to develop the habit.

But Mito didn't feed the monkeys in this manner. Her reaction to this suggestion, when I talked with her, was one of polite incredulity. Like anyone familiar with the strict hierarchy in a macaque society, Mito realizes that one cannot hand out food any way one wants. A large section of the group is low on the totem pole, and any monkey who possesses food at a time when the highest-ranking males have none is bound to get into deep trouble. Imo and her peers could therefore not have been favored too much at feeding time lest Mito put their lives in danger, which I am sure she did not want to do. To keep dominant males from making trouble, they needed to be fed first. And remember, adult males were the last ones to learn potato washing, if they learned it at all.[144]

It became clear in my discussions with Mito that the feeding of potatoes started in the forest, away from the beach. Hence, the first potato washers had to actively seek out the water. Later, when the troop became better habituated and more attracted to the open area, provisioning nearer to the beach became part of the routine. It was again on their own initiative, however, that the monkeys switched their attention from the freshwater stream to salty water. In other words, the monkeys were in charge in all of this, not Mito.[145]

The second pillar of Galef's claim was that the speed with which the troop had adopted potato washing was far too slow for social transmission. He wrote: "One probable advantage of social learning over trial-and-error learning is that social learning is more rapid than trial-and-error learning. One sign of social learning should, therefore, be relatively rapid spread of a behavior through a population."[146] If social learning would be faster than individual learning, this would indeed be an advantage. But this is not the only advantage of social learning. As long as *what* is learned is a useful behavior that an individual would not have acquired by itself, the advantage is enormous regardless of speed. In addition, is it really that slow if a habit reaches half the population in five years, and the entire population under middle age in a decade? In the larger scheme of things, this is rapid change indeed, not unlike that induced by baby boomers in our own societies.[147]

The one point to concede to Galef is that it would be hasty to conclude that imitation was involved, if imitation means the intentional copying of someone else's actions. There is no evidence for this in the Koshima records, and the Japanese investigators themselves have wisely been neutral on the mechanism of transmission. It may have been as simple as stimulus enhancement, that is, one individual joins another who washes a potato, picks up pieces dropped by the other in the water, finds the taste attractive, and is thus primed to develop the same habit.

The strongest argument for social learning is the way the habit's trajectory through the group followed peer relations and kinship ties. It can hardly be coincidental that one of the first to follow Imo's example was her mother, who was seen washing potatoes within a few months of her daughter. The mother was clumsy at first, but gradually became more skillful, suggesting that even if she caught the "idea"—the association of potatoes with water—from her daughter, she still needed to refine the washing technique on her own.

This possibility doesn't strike me as farfetched, but after the monkeys had successfully spread their habit, people couldn't resist spreading rumors. In my field, there is vigorous debate about the validity of "anecdotal" information (based on single, unique observations), but at least most of the anecdotes that we consider are firsthand accounts by experienced animal watchers. In contrast, the overly enthusiastic story about monkey telepathy and the overly skeptical one about human influence are both armchair speculations.

Still Doing It

For over a quarter of a century now, the Koshima monkeys have received sweet potatoes only a couple of times per year. The food is unceremoniously dumped onto a dry section of the beach, after which there is a free-for-all. There is no selective reinforcement, no encouragement to approach the ocean, not even a need to clean the food. This is the situation most monkeys on the island have experienced all their lives. Yet, from the old to the very young, they are still doing what the late Imo began. Persistence of habits beyond the life of the initiator is one of the characteristics of culture.

Not only did Imo set things in motion on the island, she also got the juices flowing in the international scientific community. Is culture a uniquely human capacity, or are we justified in applying the concept to animals? It would indeed have been *baka-rashi* for Imanishi and his students to have stayed with horses while such fascinating discoveries were waiting to be made nearby.

While we were preparing to leave, the island had grown unusually quiet. All of the monkeys had settled down on the beautifully shaped rocks to groom, or just sit and nap. Apart from the occasional tantrums associated with weaning—quickly resolved by a maternal embrace—the monkeys had taken on a dreamy attitude, no doubt due to the combination of a full stomach and a setting sun. Walking on the beach, I found the scene positively surrealistic. The eerie silence was reinforced by the fact that the monkey population here competes with birds over scarce foods, and the birds have lost. Apart from one diving osprey, and the ubiquitous kites high up in the air, there was not much that flew around. But for me the air was filled with history.

6

The Last Rubicon

Can Other Animals Have Culture?

"We can approximate what culture is by saying it is that which the human species has and other social species lack."

Alfred Kroeber, 1923

"The ability to transmit learned behavior from generation to generation gave mammals an overwhelming advantage in the struggle for existence."

Ralph Linton, 1936[148]

*T*he question whether animals have culture is a bit like whether chickens can fly. Compared to an albatross or falcon, perhaps not, but chickens do have wings, they do flap them, and they do get up in the trees. Imagine a world devoid of flying creatures *except* for the chicken: we would probably be mightily impressed, writing poems and songs about how we wished to be like them.

Similarly, viewed from the towering cultural heights achieved by the human race in art, cuisine, science, and political institutions, other animals seem nowhere in sight. But what if we change perspective, and don't measure them by our standards? This is what Kinji Imanishi proposed in the early 1950s by defining culture, not by technical achievements or value systems, but simply as a form of behavioral transmission that doesn't rest on genetics.[149]

We can also turn the question around, and ask how smart animals with a development that takes about five years, such as most monkeys, or up to twelve years in the apes, and even longer in elephants and whales, could possibly *not* transmit information across generations? How could they *not* pick up habits and social skills in those many years of interaction with their elders? In short, how could they *not* develop some sort of culture? The presence of culture, as defined by Imanishi, is nothing surprising. Sure, we can have long debates about *how* the information is transmitted, how similar or different it is to the human way, but *that* it occurs is entirely predictable.

Curiously, anthropologists have hardly contested the idea of culture in animals even though culture used to be the central concept of their discipline. This lack of territoriality is due to their own ferocious internal battles combined with postmodernist nihilism: culture has become a politically loaded, relativistic, messy concept that anthropologists have turned away from.[150]

In the old days, however, there was no reluctance to declare culture a uniquely human domain. The most influential early definition, provided by Edward Tylor in 1871, went as follows: "[Culture is] that complex whole which includes knowledge, belief, art, law, morals, custom, and any other capabilities and habits acquired by man as a member of society."[151] A century later, anthropologists placed culture on a pedestal. After all, Émile Durkheim had managed to seal the social domain off from biology, saying that only the social can explain the social, whereas both Sigmund Freud and Claude Lévi-Strauss had declared culture a victory over nature. The routine summary line then, and still today, is that culture is what makes us human. Thus, in a book ironically entitled *The Evolution of Culture* (1959), Leslie White simply declared: "Man and culture originated simultaneously; this by definition."[152]

With views such as these dominating one discipline, sparks should have flown when another claimed culture in animals. Yet, as said, cultural anthropologists are nowhere in sight in this debate. The chief critics are learning psychologists who, natural enough for them, have given center stage to the question of *how* cultural transmission takes place. They argue that culture requires imitation, teaching, and language, hence that the concept cannot possibly apply to other species.

To a biologist, this definition sounds contrived, however, because it puts the mechanism—the way things work—first. It is like defining an automobile as being fueled by gasoline, thus excluding electric cars. This is not the way things are normally defined in the life sciences. Respiration, for example, is defined as an exchange of gas molecules between the organism and its environment without specifying whether the process takes place through skin, lungs, or gills. And locomotion is defined as self-propelled movement, regardless of whether it involves legs, fins, wings, slithering body movements, or jet propulsion. Given this entirely functional outlook, biologists are perfectly comfortable with a definition of cultural propagation that remains agnostic about how it is accomplished.[153]

Nevertheless, the debate about whether animals possess culture now revolves largely around the question of what they learn from each other, and how. To what degree does it resemble the way we learn from each other, and are other animals as dependent on this process as we are? While keeping the last question for the next chapter, I intend here to show that, even though social transmission in animals is imperfectly understood, the process is highly variable and sometimes quite sophisticated. I will also propose a new way of looking at the motivation behind it, which has its roots in social emotions and conformist desires rather than reward and punishment. In the same way that the sushi master apprentice is said to absorb information without any reward for years, animals watch others and copy their behavior just so that they fit in and act like the rest.

Do Apes Ape?

In our chimpanzee colony at the Yerkes Primate Center in Atlanta, Georgia, infants sometimes get a finger stuck in the compound's fence. Their finger has been hooked the wrong way into the mesh and cannot be extracted by force. The adults have learned not to pull at the infant; victims always manage to free themselves eventually. In the meantime, however, the entire colony has become agitated: this is a dramatic event analogous to a wild chimpanzee getting caught in a poacher snare.

On several occasions, we have seen other apes mimic the victim's desperate situation. The last time, for example, I approached to assist but received threat barks from both the mother and the alpha male. As a result, I just stood next to the fence watching. One older juvenile came over to reconstruct the event. Looking me in the eyes, she inserted her finger into the mesh, slowly and deliberately hooking it around, and then pulling as if she, too, had gotten caught. Then two other juveniles did the same at a different location, pushing each other aside to get their fingers in the same tight spot they had selected for this game. These juveniles themselves may, long ago, have experienced the situation for real, but here their charade was prompted by what had happened to the infant.

I wonder where this behavior would fall under the usual classifications of imitation: no problem was being solved, no goal was being copied, and no reward was procured. The juveniles seemed fascinated by the infant's predicament, and their imitation seemed emotionally charged. It had an element of identification, of empathy and closeness, rather than the cool evaluation of goals and methods that the scientific literature proposes as the hallmark of imitation.

An (admittedly rather Dutch) example of the latter would be that I show an ape how to fix the inner tube on a bicycle after which the ape, given a bike, a tube, and the required tools, would do the same. But why would he? Consider the conditions that would need to be in place.

Identification: The ape would need to know me and care about me as a person, because otherwise why should he pay attention? Apes are hostile to, or at least uncomfortable with, strangers. They pay attention to them only with apprehension and suspicion, not with much desire to mimic them.

Understand the goal: The ape would need to know what a bicycle is and what it is good for. He also would need to know how hard it is to ride one with a flat tire.

Background knowledge: The ape would need to understand each manipulation, such as detecting the hole in the tube, using scissors to cut a patch, gluing the patch in place, pumping up the tube, listening for escaping air, and so on. Can one understand the workings of scissors or a pump without ever having handled them?

It has taken science a long time to realize that imitation is actually very complex. I remember the days when humanlike behavior of apes, including their use of tools, hand gestures, and facial expressions, was dismissed as *mere* imitation. "They act so much like us," people would say, "because they mimic everything we do. None of this is their own idea." We know now that, if this notion were true, if they could really adopt all of our facial expressions at will just from watching us, this would be spectacular! Imitation is seen as one of the highest cognitive feats. Think about it: how does one get from watching another individual's actions to performing the same actions for the same purpose? Imitation requires that visual input is converted into motor output, telling the body to reenact what the eyes saw.

Dogs or cats don't strike us as great imitators even though they are continuously exposed to our behavior. To understand what we are doing and what our reasons are is mostly beyond their abilities except when it comes to emotions. Thus, they seem to understand if we're alarmed by a noise at night (most likely they heard it before us), or when we're angry or anxious, but that we want to keep the house clean is—as they demonstrate every day—beyond their comprehension. Apes are light-years ahead of the average domestic pet in this regard, but even they generally don't grasp the exact reasons behind our actions. They often copy human actions without understanding. Cleaning the house is a case in point, as in the following account by Nadie Ladygina-Kohts of her adoptive chimpanzee, Yoni:

> Left to his own devices, Yoni often takes a broom or brush and tries to sweep the floor, raking the trash into a pile. However, he does it so awkwardly and inefficiently, due to a lack of direction, that he rather spreads the trash over the floor than gather it, and the floor is never clean as a result of that. Yoni even moves the furniture, as is done during a

clean-up, although he often does not sweep the floor at the freed spot.[154]

There are many such examples, all of which concern apes who love to be with people, following their every move. For example, Anne Russon recently reviewed instances of imitation of typical human behavior by rehabilitant orangutans in Borneo's Tanjung Putting National Park, including sawing wood, hammering nails, putting on a T-shirt, shading the eyes with a hand, and stringing up a hammock. Here is an example of personal hygiene copied from people:

> Davida, an adolescent female rehabilitant, came to the bunk house porch one morning about the time when people came out to wash. A visitor gave her a toothbrush with toothpaste on it. Davida and others had in the past stolen toothbrushes and toothpaste but rarely both together. She put the brush in her mouth, nibbled the paste, then brushed: with the brush in her fist, she inserted its bristle end in one side of her mouth, closed her lips around it and scrubbed the bristles back and forth. When she finished brushing, she climbed onto the porch railing, spit the used toothpaste over the railing onto the ground, then moved off. Her technique of brushing, including spitting used toothpaste over the rail onto the ground, was identical to the standardized technique used by camp visitors at that house.[155]

Whether Davida had any idea *why* people brush their teeth remains unclear, but there can be no question that, like Yoni, she was an excellent observer and copier. Despite the overwhelming anecdotal evidence that it is not for nothing that the word "ape" has in many languages inspired a verb meaning "to imitate," scientists have not always been successful getting apes to ape in the laboratory. The typical approach has been to let them watch a simple task demonstrated by a white-coated, barely familiar human experimenter, who operates with neutral composure. Experimenters aren't supposed to influence their subjects, hence schmoozing, petting, or other niceties are discouraged. But since to influence the subject is the whole point of these experiments, one could argue that these procedures violate the most basic requirement for imitation, which is that the ape feels close to the model and identifies with him or her. Moreover, the outcome is customarily compared to tests in which human children get exactly the same treatment. But children are human, which means that the experimenter belongs to their own species, acting in a way perfectly comfortable to the child. No wonder children do better.

There is an old rule in science according to which absence of evidence is not the same as evidence of absence. Yet some scientists have gotten great mileage out of negative evidence, taking the "failure" of apes in these kinds of comparisons with children to mean that imitation is restricted to our species. Before drawing any such conclusions, however, shouldn't we level the playing field by either giving children the same

Prince Chim, a young bonobo, adopts the poise of a serious student. Even though imitation in animals has become a controversial topic, there can be little doubt that apes spontaneously copy whatever they see others do. Robert Yerkes wrote that again and again Chim was seen to take a book and turn the pages carefully and neatly one by one, as if he wanted to discover what humans found so interesting about this activity. (Photograph by Robert Yerkes, taken in 1923, with permission of the Yerkes Primate Center).

species barrier faced by the apes, or else removing it for the apes? One could, for example, test children with ape models to see whether they would still be better imitators. Another possibility would be to present apes with a model of their own kind to see how they compare to children following a human model. Finally, one could present a human model to apes who are completely and fully familiar with our species.

The latter is the most practical alternative, and as soon as it is implemented the differences between apes and children vanish. In one such comparison, by psychologist Michael Tomasello and co-workers, apes raised like human children in a language laboratory turned out to be as good at imitating people as two-year-old children, whereas apes raised by their own kind did rather poorly.

The surprise here is how the investigators interpreted this result. Instead of concluding that apes are a match for young children when both are equally familiar with the model, they conclude that human-reared apes are special. They consider these apes *enculturated*, meaning that the enriching, stimulating context of the human environment has brought out capacities that these animals normally don't have. By claiming that apes can't imitate unless they have benefited from humanity's shining light, they leave the human-ape divide intact.[156]

But why should apes have evolved a cognitive potential that they don't use or need in their natural life? Evolution is rarely wasteful. I prefer the simpler view that apes are born imitators, that this talent serves them in their natural social life, but that they prefer to imitate the species that has raised them. Under most circumstances this will be their own kind, but if reared by another species, they will imitate that one as well.[157]

Language-trained apes often give the impression that they regard themselves as almost human, such as when, while sorting pictures of humans from those of other animals, they put their own portrait on the human pile. They obviously sympathize with the people that surround them: they want to fit in, and to be like them.[158] Being raised by another species creates familiarity with its communication and interest in its actions. As a result, rather than having been lifted to unprecedented cognitive levels, apes raised by people have become ideal test subjects simply because they are willing to pay attention to psychologists. Their humanlike ways remind me of an example by Charles Darwin in *The Descent of Man*, which—even if unconfirmed—creates an irresistible image of one species struggling to be like another: "[T]here is reason to believe that puppies nursed by cats sometimes learn to lick their feet and thus to clean their faces."[159]

The Urge to Be Like Others

In the minds of some scientists, "monkey see, monkey do" has been transformed from something obvious, almost stupid, to a miraculous achievement: the Holy Grail of our magnificent cultural abilities. But, as we have seen, the way imitation is being defined and evaluated disadvantages apes who fail to pay attention to the species that does the testing.

I am not saying that apes are capable of copying as complex a series of manipulations as repairing a bicycle. I used this example precisely because what may seem simple to us is often far too complex for the ape. Generally, imitation concerns novel acts or solutions that the animal is on the brink of discovering on its own; seeing someone else do it is the last push they need. Hence, imitation matters especially in domains in which the animals are already strong. From watching another, an ape may learn the use of a twig or stone to obtain food, or the correct way of holding a newborn. Those are the sorts of things they are interested in. Most human actions, in contrast, mean nothing to them and stay forever incomprehensible.

There is a veritable war going on in academia about imitation, with definitions up for grabs, and new levels, from the simplest to the smartest, being introduced on a daily basis. One scientist's "imitation" is another's "emulation," and a third's "facilitation."[160] The central idea remains that one individual adopts another's behavior, which it most likely would never have done without exposure to the other. Apart from this common ground, there is precious little agreement about the hows and whys of the process.

What I see in the daily life of chimpanzees is a tendency for youngsters, especially, to act like others in the group, namely, their elders. For example, our dominant male, Socko, makes a spectacular charging display, slapping the ground loudly with his hands, kicking a few empty barrels, and throwing a piece of wood around. He may do so for ten minutes, during which the entire group watches him with apprehension, while some of the other adult males perform their own displays at a safe distance. When everything has calmed down, often many minutes later, a young male, only three or four years old, will put up his hair and charge at one of the barrels, kicking it in the same way as the boss had done.

As another example, one of the chimpanzees has an injured finger, and walks around leaning on a bent wrist instead of his knuckles. During the same period that he's hobbling around in this odd way, all youngsters in the group walk around on their wrists, too. They don't necessarily do so when the injured male is around; they do it everywhere and all the time. It has become a fashion. Sometimes they follow the target of their imitation. At the Arnhem Zoo, we had a female named Krom (meaning "crooked") because of her hunched-up walk, who was often followed by a line of juveniles with the same pathetic carriage.

The word "fashion" was first used in relation to chimpanzees by Wolfgang Köhler (1925), whose apes invented new games all the time. The following account gives an impression of how group-oriented and act-like-others a species they are:

> The whole group of chimpanzees sometimes combined in elaborate motion patterns. For instance, two would wrestle and tumble about playing near some post; soon their movements would become more regular and tend to describe a circle round the post as a center. One after another, the rest of the group would approach, join the two, and finally they march in orderly fashion in single file round and round the post. The character of their movements changes; they no longer walk, they trot, and as rule with special emphasis on one foot, while the other steps lightly; thus a rough approximate rhythm develops, and

they tend to "keep time" with one another. They wag their heads in time to the steps of their "dance" and appear full of eager enjoyment of their primitive game.[161]

To bring these tendencies to bear on the issue of culture, we need only look at how chimpanzees learn to crack oil-palm nuts. According to field-workers, the expertise of their animals far exceeds that of any human who tries it for the first time. It takes many years of practice to place one of the hardest nuts in the world on a level surface, find a good-sized hammer stone, and hit the nut with the right speed so as to crack it. It is the most complex tool-use task known from the field, involving both hands, two tools, and exact coordination.

A Japanese primatologist, Tetsuro Matsuzawa, carefully tracked the development of this skill in wild chimpanzees at Bossou, Guinea. Young chimpanzees join the rest of the group at the cracking "factory": a location where the apes gather nuts around anvil stones, pick up hammer stones, and fill the jungle with a steady rhythm of banging noises. Youngsters hang around with the hardworking adults, occasionally pilfering nuts and stones from them to try things out for themselves. They also get a good deal of food from their mothers, who share the kernels of cracked nuts. In this way they learn about the edibility of nuts, and with time also about the connection with stones.

At first, infants are seen handling single objects. They play with a nut or a stone, but do nothing with them together. At the next stage, infants begin to randomly combine objects. They put nuts on anvil stones, or they push stones and nuts together. They also spend quite a bit of time hitting nuts with a hand, or stamping them hard with a foot, which of course does nothing to open them. Only after three years of futile efforts do they finally begin to coordinate multiple actions to crack a nut with a pair of stones, using one as anvil and the other as hammer. They still need a great deal more coordination and refinement, and so it is only when they are six or seven years old that their skills begin to approach those of adults.[162]

What does this tell us? Here we have young apes whose actions gradually begin to resemble those of their parents without ever having been rewarded. Since they fail continuously for at least three years on end, the incentive for their imitation cannot be in its payoffs. They may even experience negative consequences, such as smashed fingers, or the frustration of knowing that there's food inside the nut while not being able to get to it. What keeps them going?

This question comes up because it has been suggested that imitation never exists for its own sake, but is invariably strengthened or weakened by what it delivers. Bennett Galef, one of the experimental psychologists who, as discussed earlier, is skeptical about animal culture, described the need for reinforcement as follows: "In my view, although imitation might introduce some novel behavior into the repertoire of members of a population, through time (probably counted in days) this behavioral novelty would be maintained, modified, or extinguished depending on its effectiveness in acquiring rewards."[163] This sounds logical enough, but is it consistent with the facts? The field research on nut cracking rather shows that young chimpanzees perform an unrewarded activity, copied from their elders, for *over one thousand days* without ever slowing down. Could it be that when it comes to cultural transmission, the traditional emphasis of learning psychologists on tangible incentives is misplaced? Perhaps the copying of others is more like a drive, that is, it reinforces itself.

This would quite simply mean that social learning is *socially* motivated. A young chimpanzee, for example, feels close to mother, identifies with her, and expresses this closeness by watching all her moves and doing everything like her. Young chimpanzees are constantly in search of role models for infant care, feeding techniques, dominance displays, sexual intercourse, and so on. It is this social orientation that feeds their mimicry. Only when the apprentice cracker has achieved enough dexterity and strength to actually open a nut does food enter the picture. This doesn't mean that before this moment nut cracking was not goal-oriented: I rather look at it as a *shifting* of goals. At first, there is the orientation to the mother, and the desire to act like her. In the process, almost by accident, the second goal—to feed on tasty kernels—emerges and gradually takes over from the first.

The same may be true of monkey imitation. Michael Huffman, an American who has worked for twenty years on

Arashiyama, a mountain overlooking Kyoto, reports the curious habit of Japanese macaques of rubbing stones together. The monkeys often come down from the mountain to a flat, open area where they receive food from park wardens and tourists. Daily, one sees them collect handfuls of pebbles or small rocks. They carry these to a quiet spot, where they rub or strike them together or spread them out in front of them, scattering them, gathering them up again, and so on. When I first saw this, it looked as if they were trying to make fire, but this is of course again the human goal-seeking mind at work. Young monkeys learn this totally useless activity from peers, siblings, and their mothers, resulting in a widespread tradition within this particular troop. As Huffman notes, "It is likely that the infant is first exposed *in utero* to the 'click-clacking' sounds of stones as it mother plays, and then exposed visually to stone handling as one of the first activities it sees after birth, when its eyes begin to focus on objects around it."[164]

Exactly how the monkeys learn from each other remains a question, but it is obvious that young monkeys copy the stone handling without any reward other than perhaps the noise associated with it. Nevertheless, there are no signs of diminished enthusiasm; for decades, the activity has consistently been passed on to each and every infant born in the troop. If there is any case in the primate literature that refutes Galef's assumption that imitation requires rewards, it is this curious behavior of the monkeys on Arashiyama.

Let me therefore propose an alternative view, which is that primate social learning stems from conformism—an urge to belong and fit in. To give this process a name and emphasize that it favors certain social models, such as mothers and peers, I will use the acronym BIOL, which stands for Bonding- and Identification-based Observational Learning. Instead of being dependent on tangible benefits, such as food, BIOL is a form of learning born out of the desire to be like others. Certain social models are copied in an often playful, imperfect, and exploratory fashion. Whether or not this translates into rewards is secondary.

For this reason, BIOL is sometimes more style than substance. A good example is the N-family of rhesus monkeys that I studied at the Wisconsin Primate Center. This matriline was headed by the matriarch, Nose, all of whose offspring had been given names starting with the same letter, such as Nuts, Noodle, Napkin, Nina, and so on. Nose had developed the habit of drinking from a water basin by dipping her entire underarm into it, then licking her hand and the hair on her arm. It was amusing to see how her offspring, and later her grandchildren, followed the same technique. No other monkeys in the troop, or any other rhesus troop that I knew, drank like this. Unless one wants to propose a gene for drinking style— which would be a stretch—the tradition must have been transmitted through observation. And again, rewards played no role, since the N-family did not obtain anything that other monkeys were unable to obtain.

The self-reinforcing quality of BIOL has been overlooked. We are such law-of-effect creatures that we have trouble looking at imitation from a purely socioemotional perspective. We see purpose everywhere, and if we don't see it we feel there must be something wrong. But even if imitation is mainly socially driven, aimed at emulating favored models, the end result remains that habits and techniques spread through a population. It is at this level that the payoffs occur. The individual does not need to realize this, such as when a monkey learns to fear snakes without ever having been bitten or when a playing chimpanzee begins to combine stones with nuts. Or think about the song learning of birds: no one has ever suggested immediate rewards for their imitation. Insofar as conformist tendencies contribute to survival, they will be selected for. Indeed, the argument has been made that the desire to act like others and the ability to copy others evolved in tandem, allowing individuals to take full advantage of the knowledge and adaptive habits around them.[165]

The Tortoise and the Hare

When we look back at the three earlier requirements for imitation—identification with the model, understanding of its goals, and presence of basic pre-knowledge—it is evident that the first and last are common in primate societies, underlying much of the observed imitation. But whether there is much understanding of the model's goals remains questionable. The imitation of monkeys and apes may be less goal-oriented than what we humans achieve at our best moments. This is not to say, however, that most of our imitation—a boy walking like his father, a teenager talking like her friends—isn't exactly as socioemotional as the BIOL of other primates.

The critical question now is whether such a deficit—if that is what it is—in the imitative capacities of monkeys and apes is reason enough to ban them from the cultural domain. This is a favorite argument of some skeptics, the most extreme position being that of psychologist David Premack, who, together with his wife, Ann, claims that imitation is a spontaneous activity of the human child but not of the chimpanzee. The Premacks go so far as to say that "in non-human primates, imitation plays no role in the transmission of information, whereas in humans it plays a major role."[166]

Curiously, in the same article they contradict this statement right away by discussing striking anecdotes of their own chimpanzee, Sarah, who apparently is a great imitator. Following a logic similar to Tomasello's concept of enculturation, they ascribe Sarah's brilliance to the human environment, saying that "field observations are poor forecasters of the potential behavioral complexity of the chimpanzee." Hence the idea is that human-reared apes operate on a different mental plane than wild ones, having benefited from human cultural stimulation.[167]

The challenge for those defending this position is to explain why apes, and not other animals that are extensively exposed to people, such as dogs, can be pushed to such great heights. What makes apes so exquisitely educable? Any sensible theory will, in my opinion, propose that they are sensitive to human enculturation precisely because cultural influences abound in their natural environment. They are educable because they need to be. If so, the enculturation idea supports rather than contradicts the possibility of ape culture.

Also, it is not as if Sarah is the first ape for whom imitation has been reported. Why would these observations, made up close in the laboratory, be more relevant than similar stories of imitation by apes under natural or naturalistic conditions? I am unconvinced that what an anonymous youngster does at Bossou, playing with stones and nuts until he can do what the adults around him are doing, is any less impressive than Sarah's laying a banana peel like a hat on top of a portrait after having seen a person wearing a hat. Both apes try to re-create what they have seen.

The Premacks go on to remark that chimpanzees have been studied in the field for perhaps a hundred years but that nothing important has ever changed in their behavior. This lack of change is seen as an argument against chimpanzee culture. But couldn't exactly the same argument be used against human culture? Conservation and inertia are as much part of culture as innovation. Until recently, some peoples still lived in the Stone Age, and humanity's prehistoric record shows periods of hundreds of thousands of years without any appreciable change in tool manifacture. Chimpanzees, in contrast, have been carefully studied in the field only for about forty years.

Speed is never what defines a process: the hare and the tortoise both get from A to B. Even if it were to take chimpanzees a million years to develop and transmit the habit of waving flags when the moon is full, I would still categorize this behavior as a cultural tradition. Here is what a classical anthropological text, by Alfred Kroeber, says about the speed of change:

> One of the most widely held preconceptions is that culture is progressive. "The progress of civilization" is a familiar phrase—almost a trite one. Simple or primitive peoples are labeled "unprogressive." The implied picture is a of a continuous moving forward and onward.
>
> Actually, the idea of progress is itself a cultural phenomenon of some interest. Strange as it may seem to us, most of humanity during most of its history was not imbued at all with the idea. An essentially static world, a nearly static mankind, were most likely to be taken for granted. If there was any notion of alteration, a deterioration from the golden age of the beginnings was as frequently believed in as an advance.[168]

Solid Ground

In 1999, Stephen Jay Gould commented in the *New York Times* that humanity has an unfortunate tendency to erect "golden barriers" that set us apart from the rest of the animal kingdom. He was reacting to reports in *Nature* about cultural primatology.[169] Golden barriers are bound to be rejected, argued Gould, using the reports as proof of such rejection with respect to humanity's purported cultural uniqueness. Within days there were letters by distressed readers in the *Times* about how what apes do is "instinctive," whereas what humans do is "conscious," and how only humans possess a free will.

In the minds of many people, culture is associated with freedom. They believe with Ashley Montagu that "man has no instincts."[170] But doesn't culture restrict our freedom as much (or as little) as biology? And where do our cultural capacities come from? Don't they spring from the same source as the so-called instincts? How could one possibly, with learning and cognition so conspicuous in the lives of chimpanzees, call everything they do instinctive? Whereas we can fully expect that definitions of culture will keep changing so as to keep the apes out, the proposals heard thus far seem insufficient to do so.[171] Culture requires imitation and teaching? It needs to be fast? Apart from the already mentioned problem that mechanisms are poor criteria, the question becomes again how human culture would fare. The following items would probably not be considered cultural if we were to follow the definition of experimental psychologists, which center on imitative problem solving:

- Clothing, ornamentation, taste in colors
- Religion
- Cuisine and food preferences
- Music, art, and dance
- Social styles, such as egalitarian versus hierarchical, or polite versus rude societies

None of these items has much to do with problem solving. If this is the touchstone, neither the Spanish flamenco nor the Scottish kilt nor Chinese food would be cultural. Needless to say, this is not how we usually employ the term. Owing to my wife's and my disparate backgrounds, I know subtle yet pervasive cultural differences firsthand. Even after decades together, we still don't agree on the right time for dinner, we have different tastes in color, and we disagree about what to discuss in public, and how. Cultural imprinting often takes place through sheer exposure and force of habit. If you bow all your life for parents, teachers, and other higher-ups, how can you ever get used to a culture where people stand upright, shake hands, and call each other by their first names? This is culture at work in our daily lives: a broad set of influences on customs, habits, tastes, attitudes, and sensitivities.

I therefore much prefer Imanishi's definition of culture as nongenetic behavioral transmission. With this definition, we can include both the entire rich tapestry of human culture and behavior transmitted by other animals in ways that overlap with but are not necessarily identical to ours.

Remarkably, despite his reflection in the epigraph to this chapter, and after having produced the most comprehensive review of anthropological definitions of culture, Alfred Kroeber didn't rule out animal culture as a possibility. He didn't necessarily cling to the human standards of language, values, and full-blown imitation as part of culture. Hearing of the "dance" of Köhler's chimpanzees, Kroeber speculated that if an ape developed a new dance step, or a new posture, and if

these acts were then picked up by others, became standard-
ized, and survived beyond the inventor's generation, "we
could legitimately feel that we were on solid ground of an ape
culture."[172]

By accepting a broad, inclusive definition we can make the
necessary connection between human and animal culture
without in any way devaluing the former's achievements. The
fact that primates sometimes duplicate behavior, such as the
rubbing together of stones or a special drinking technique,
that confers zero advantages is extremely telling. It teaches
that cultural learning is not about rewards, but about fitting in.
Identification with others and a desire to conform are tendencies we easily recognize in ourselves, as reflected in concepts
such as "peer pressure" and "role model." We can now assume
these tendencies, which underlie all forms of cultural transmission, to be far older than our own tenure on this planet.

The Nutcracker Suite

Reliance on Culture in Nature

"Certain kinds of information can only be transmitted by behavioral means. If the transmission of this kind of information is adaptive, then there would be strong selection pressure for culture."

John Bonner, 1980

"Chimpanzees would not greatly miss their so-called tools. They would find other things to eat, get by without medicines, not perish for want of leaf-sponges."

Barry Allen, 1997

Nut cracking is such a noisy business that a French colonist, in Ivory Coast, once speculated that an unknown jungle tribe must be forging iron. A Portuguese Jesuit in Sierra Leone got closer to the truth, writing, in the early

239

1600s, that he had heard how chimpanzees collected handfuls of palm nuts in the forest to break them open with a stone.[173] Obviously, this would have been hard to do with a single stone. We had to wait until 1951 for the first eyewitness account, published as a half-page note in a scientific journal by Harry Beatty, an American expatriate to Liberia, who probably had no idea of the momentous importance of his discovery:

> At last, I arrived at the edge of the dark, matted vegetation and glanced along the path before me. It was deserted, but I noticed a shell heap twenty-five feet away. . . . A moment later an adult male chimp came ambling around a bend, bearing an armful of dried palm nuts, supporting himself on his right knuckles. He reached the rock, sat down clumsily and proceeded to select a nut. He then picked up a chunk of rock and pounded the nut which had been placed on the flat surfaced rock. I watched this procedure for several minutes until an alarm note from the rear sent the troop scurrying off.[174]

The first primatologist to describe the same behavior was Yukimaru Sugiyama, of Kyoto University. Together with a Guinean colleague, Jeremy Koman, Sugiyama found large stones accompanied by smaller ones and empty nut shells in the forest of Bossou, Guinea. In the surface of the larger stones, called platforms, they found shallow cavities indicating that they had been in use for a very long time. In 1979, the two field-workers published a report on twenty-nine cracking sites in the forest and three direct observations of apes placing nuts one by one on the platform and pounding them with smaller stones. These hammer stones weighed under one kilogram but were nevertheless "a little too heavy for the authors' use." Because all of this took place in a dark forest, no photographs could be taken of the chimpanzees actually doing it. For many years, others questioned the observations.[175]

Villagers around Bossou use the same technique to open the same nuts. Their cracking tools are indistinguishable from those of the chimpanzees except for generally lighter hammer stones. Human and ape cracking sites are hard to confuse, though, as people rarely venture into the forest and chimpanzees never enter the village. It remains unclear whether the technique could have been transmitted between people and apes, and if so, in which direction. The discoverers go on to observe: "This kind of tool must be almost identical to those by earliest man. Indeed, if they had been found at excavation sites, archeologists almost certainly would have judged them as human stone tools from the unnatural cavities on the pebble tool and platform stone surfaces."

This remark explains the skeptical reception of their findings. Stone tools of any kind used to be attributed without question to humans or their direct ancestors; hence, it was disturbing to hear that there might be other possible producers. We now know much more about nut cracking by chimpanzees, which, because it involves the use of two tools simultaneously in precise coordination, is considered their most complex sub-sistence technology. Lack of dexterity will result in hurt fingers, escaping nuts, or pulverized contents. First-time human crack-ers often don't succeed at all. Here is how two primatologists, Christophe Boesch and Hedwige Boesch-Achermann, de-scribe nut cracking deep in the Taï Forest, in Ivory Coast, where chimpanzees use extremely heavy stones to crack the world's toughest nuts, each requiring extraordinary force and an average of thirty-three blows before they give up their prize:

> To extract the four kernels from inside a panda nut, a chimp must use a hammer with extreme precision. Time and time again, we have been impressed to see a chim-panzee raise a ten-kilogram stone above its head, strike a nut with ten or more powerful blows, and then, using the same hammer, switch to delicate little taps from a height of only a decimeter. To finish the job, the chimps often break off a small piece of twig and use it to extract the tiny fragments of kernel from the shell.[176]

But Is It Tool Use,
and Do They Need It?

That chimpanzees use such fine coordination, that their cracking sites could be mistaken for those of lithic people, that only certain wild communities know how to crack nuts, that females are better at it than males, and that it takes youngsters many years to acquire the skill—all of this has been learned in the last few decades against the backdrop of earlier claims about tool use as a uniquely human capacity.[177] Or, if tools by themselves weren't exclusively human, then at least the manufacturing of tools was. The classical experiments by Wolfgang Köhler on chimpanzees, combined with Jane Goodall's pioneering observations in the field—where she saw chimpanzees modify branches to make them suitable for termiting or chew leaves before using them as a sponge—effectively eroded these claims.[178]

Yet even today objections can be heard. Incredible as it may sound, tools can be defined in such a way that chimpanzees must be using something else. Some scholars have emphasized that human technology is embedded in role complementarity, symbols, cooperative production, and education. In their opinion, the "tool" label isn't deserved if some or all of these elements are missing, such as when a chimpanzee cracks nuts with rocks, or, I suspect, when a farmer picks his teeth with a twig.[179] To draw a clear line is all the more important, according to philosopher Barry Allen, because chimpanzees don't really *need* their so-called tools. All this talk of ape tools is based on feeble analogies, says Allen, that have nothing to do with our own dependence on technology. In a melodramatic passage, he goes on to bemoan the damage already done to our self-perception:

> Belief in the chimpanzee's tool has already proved misleading in the history and philosophy of technology. If it is a scientific fact (as a litany of authorities attest) that chim-

panzees use tools (real tools, like ours), and if those tools are superfluous, things that chimps could get along without, a possible conclusion is that humans, too, do not really need tools. Technical culture is habitual among us but not an animal necessity, conceivably even superfluous.[180]

Is the human self-image really that fragile? We all realize how essential tools are in our daily lives—we can't even do simple kitchen tasks without them—so why would we, upon hearing that apes in the forest use tools, rush to the conclusion that we don't really need ours? Does the one follow from the other? Apart from this, why would anyone even suggest that apes could easily survive without their tools? It is only from a philosopher's armchair that it is conceivable that wild chimpanzees sit there pounding and pounding, thirty-three blows per nut, for generation after generation, for no good reason at all.[181]

During peak season at some sites, chimpanzees spend fifteen percent of their waking hours fishing for termites. At other sites, they devote an equal amount of time to cracking nuts. Careful measurements show that they may gain nine times as many kilocalories of energy from this activity than they invest in it.[182] Moreover, when Gen Yamakoshi collected information on nut cracking at Bossou, he found nuts to be a keystone resource: they serve as fallback foods when the main source of nutrition—seasonal fruits—are scarce. During the rainy season there are fewer fruits in the trees, which forces the chimpanzees to switch from their usual easy pickings to hard work, devoting almost twenty percent of their foraging time to resources that are inaccessible without tools. Their main fallback foods are palm nuts, which they crack with stones, and palm pith, which they obtain through so-called pestle-pounding. High up in a tree, a chimpanzee stands bipedally at the edge of the tree crown, pounding the top with a leaf stalk, thus creating a deep hole from which fiber and sap can be collected.

These two kinds of food, which are available throughout the year, are exploited whenever fruit production in the forest drops to a low. So, rather than being a leisure activity or hobby, tool technology seems a critical skill. Given that even small differences in food intake are known to affect health, survival, and reproduction, chimpanzee communities with the right technologies enjoy an advantage. Yamakoshi attributes the relatively high reproduction rate of the Bossou population to its effective exploitation of available foods. Naturally, he argues that if indispensability of tools for survival is part of the definition of culture, his chimpanzees should be let in on it.[183]

Allen's logic that tools must be essential for us, since all human cultures have them, applies equally to the chimpanzee. We know of no wild chimpanzee population without tools; hence, claims of a major difference with the human condition are shaky. It remains possible, even likely, that not each and every known chimpanzee technology contributes to survival. But the same can be said of our tools, which range from the essential to the trivial. We will return to this issue later on, because the claim that apes don't really need their tools, that they use them as toys or gadgets, is a transparent attempt at trivializing what they do. It is part of the myth that we humans have broken with our nature, whereas animals are still stuck with theirs.

Clasping Hands

In 1992, I first saw two chimpanzees at the Yerkes Primate Center's Field Station clasp their hands together. The two were sitting in a climbing frame grooming each other when all of a sudden one female, Georgia, took the hand of another, the older Peony, lifting both of their hands high into the air. They thus sat in a perfectly symmetrical A-frame posture, each with their free hand grooming the pit of the other's lifted arm.

I recognized the posture from an old article, published in 1978 by two fellow chimpwatchers, William McGrew and Caroline Tutin.[184] The article had stayed with me because it was the first description of wild chimpanzee behavior for which I knew absolutely no equivalent in any captive group. Of course, there are other things wild apes do, such as foraging in the trees or fighting with neighbors, that don't occur at zoos or research centers, but that is because conditions don't permit it. All other typical chimpanzee behavior, especially in relation to their socioemotional life, is highly observable. The opposite is also true: behavior first recognized in captivity is sometimes later seen in nature. For example, inspired by my studies at the Arnhem Zoo on reconciliations after fights, there are currently at least three ongoing projects documenting this process in the field. In general, I believe that chimpanzees face, in their social lives at least, basically the same sort of problems in the forest as under enlightened captive conditions. They arrive at the same solutions, whether it be with regard to mother-offspring relations, power politics, or the mutual exchange of favors. Also, their communication, such as vocalizations and facial expressions, is essentially the same under both conditions.

Because of this fundamental similarity, I found it intriguing that wild chimpanzees groomed each other in such an unusual way. But not all chimpanzees do, and this is precisely the point that McGrew and Tutin made in their article. The two investigators were extensively familiar with chimpanzee behavior in Gombe National Park in Tanzania before they visited a different study site, in the Mahale Mountains. There was no reason for them to expect striking differences between the two communities, since they concern the same subspecies of chimpanzee on the same eastern shore of Lake Tanganyika, only 170 kilometers apart. Nevertheless, the Mahale chimpanzees frequently engaged in handclasp grooming, whereas this behavior remains totally unknown in Gombe.

The Mahale research team, led by Toshisada Nishida, was of course familiar with handclasp grooming, but until then they had not realized that they had a unique behavior on their hands. This is why it is so important that field-workers visit each other's sites. If one knows only one community, it is easy to think that its behavior is typical. Based on their visit, McGrew and Tutin were the first to seriously question the assumption of "typical" chimpanzee behavior. They boldly introduced the term *social custom* to describe the Mahale grooming handclasp.

With this as background, it is easy to see why I was thrilled to discover the very same custom in my own outdoor-housed group of twenty chimpanzees, which made it the first (and thus far only) captive community in the world with confirmed handclasp grooming. The great advantage was that the custom seemed to be in its early stages. Rather than encountering a wild community of chimpanzees that has been doing it for hundreds, perhaps thousands of years, here we had a group in which the behavior was initially extremely rare (only a dozen instances were seen in the entire first year of daily observation) and always initiated by the same individual, Georgia, the presumed inventor.

Georgia is affectionately known as our troublemaker. Born to one of the group's older females, Borie, she loves to stir things up, for example, by starting a screaming match with another female and then withdrawing when the heavyweights get involved. She is the most possessive ape I have ever seen in our food-sharing tests, in which we throw large bundles of branches into the compound. Whereas normally the chimpanzees gather around the possessor of a bundle to beg for a share, if Georgia is the possessor they don't even try, since it is

not worth the hassle. So Georgia is a special character, whom
I have known since she was little, because about six years before I accepted a position at Yerkes, I had already made a
spring visit for a special project. Even though at the time I
conducted hundreds of hours of observation, I didn't notice a
single grooming handclasp.

Over the years, Mike Seres, my assistant, and I closely followed the spread of the pattern from an occasional handclasp
between Georgia and her adult sister or Borie, to handclasps
among her family members without Georgia's involvement.
Later still, the pattern was also seen among other adults. In
1996, after we removed Georgia for management reasons,
handclasping continued. The custom thus exhibited one central characteristic of culture: independence of the originator.
When Georgia was returned to the group, years later, the
handclasp had become well established in all adults, both
males and females.[185]

The transmission was slow, taking several years, even
though the learning process seemed simple. After all, the
chimpanzees did not need to learn it from watching others.
Most likely, they were first shaped into a handclasp posture by
another chimpanzee, who took their hand and moved it up in
the air. Nishida compared this to the way some psychologists
"mold" gestures of American Sign Language in their apes by
bending or moving their hand or arm. Chimpanzees are very
sensitive to what others want them to do, such as when one
grooming partner decides to move to the shade and literally

takes the other by the hand, leading him or her to the desired spot. One will never see such gentle, understanding coordination among monkeys, and it is probably this compliance with others that allows for the dissemination of the handclasp. After having engaged in a number of handclasps, an individual may re-create the same kinesthetic experience with others. Having documented hundreds of handclasps, and having watched a gradual increase in duration of the posture from less than a minute at a time to sometimes over three minutes, we concluded that the spreading was cultural: it traveled along social lines and progressed to include all adult members of the community. It became a group-defining characteristic.

As with secret handshakes in our own societies, the handclasp may even have become a *symbol* of group membership. I felt this most distinctly when Socko, then a fast-growing adolescent male, returned to the group after an absence of several months. He had battled with the alpha male, Jimoh, and had been removed for treatment of injuries. Upon his return, a pandemonium erupted in which Jimoh tried to attack Socko yet again, but the females jumped to his defense. Fortunately, nothing serious happened, and hours later the entire group gathered on a climber to groom all of the tensions away. The grooming cluster was unusually tight, and I saw Socko move from one partner to another. He was clearly at the center of attention. With each handclasp—and there were many—it was as if the group was confirming that he was one of them, a male who belonged.

The "Bronx Cheer" and
Other Local Variants

The preceding descriptions of nut cracking and handclasp grooming represent merely two of many cultural variants in chimpanzees, one with unmistakable survival value, the other with no obvious benefits. New variants are being discovered almost monthly: the charting of chimpanzee culture has obviously only just begun.

Much of this work is being done in the field, but studies at zoos have a special place, since the spreading of behavior can be followed in greater detail. In another example, I documented a group of bonobos at the San Diego Zoo that during grooming customarily clapped their hands or feet together, or tapped their chests with their hands. One bonobo would sit down in front of another, clap her hands a couple of times, then start grooming the other's face, alternating this activity with more handclapping. This makes the San Diego Zoo the only place in the world where one can actually *hear* the apes groom. When new individuals were added, they were said to have picked up the habit in about two years. Since San Diego Zoo has loaned bonobos to other zoos, it would be instructive to follow up on them and see whether they are disseminating the pattern.[186]

The same bonobos developed special play activities, such as blindman's buff and a game that I called "funny faces." In the first, a juvenile would place an entire arm over his face or poke two thumbs into his eyes and walk around like this high up in a climbing frame, sometimes almost losing its balance or bumping into objects. It was a solitary game, but once one of the juveniles started it, others often followed. The youngest among them didn't really dare and usually walked around with just one thumb in one eye. The other game consisted of pulling weird faces, which were certainly not part of the species-typical repertoire, with sucked-in cheeks or chewing mouth movements. This game was played by all juveniles. They did not aim these faces at each other, but just sat around grimacing for no reason at all. Since I have never seen anything resembling these games in other bonobos, nor in chimpanzees for that matter, I assume that they are unique innovations of the San Diego colony.[187]

At Yerkes, we have one chimpanzee group in which many apes loudly splutter their lips during grooming, a behavior rare in some other groups. And a second group, the same one that also does the grooming handclasp, has developed fire-ant fishing. Ant nests in the compound itself never survive the destruction by chimpanzees, but if ants develop a red clay mound outside the fence, the chimpanzees gather long twigs from their enclosure. They poke them through the fence and let the ants walk onto them, after which they eat them off, much like wild chimpanzees do. This is a group activity that has never once been noticed in our other chimpanzees.

In another project by Andrew Marshall and co-workers, the typical, slowly rising hoo-hoo-hoo calls of chimpanzees, known as pant-hoots, were compared between two American zoos. Sonograms showed that adult males at each zoo had a distinct pant-hoot sequence in terms of the number of build-up calls and the duration of the climax yell. Males at each zoo sounded very much like each other but unlike the males at the other zoo. This is remarkable since at each place the males had various origins and hence were unrelated. They must have gone through a process of cultural convergence, resulting in a uniform local dialect. Indeed, one male developed a "Bronx Cheer" variant, which integrated a raspberry-like sound in his hooting. This unique call style was picked up by five other males at his zoo but was never heard at the other zoo.[188]

However instructive all of these studies may be in showing how apes naturally develop group-typical signals and customs, it is especially from the field that we are receiving a steady stream of reports on cultural variants, many concerning tool use. Thus, chimpanzees at Bossou, in Guinea, have been seen to arrange leaves on the ground so as to avoid sitting directly on wet soil. It is unclear how long the Bossou chimpanzees have been using these so-called "leaf cushions."[189] At a different site, in Sierra Leone, chimpanzees enter kapok trees to collect fruits, but these trees are full of sharp thorns that make it very painful for them to move around in the canopy. The apes have a unique solution that allows them to spend more time in these trees than apes at any other site. Breaking off small, smooth branches, they place these over the thorns and use them to step or sit on. Sometimes an ape holds a stepping-stick between his greater and lesser toes, walking around with it "attached" to his foot. He thus has effectively provided himself with a protective sole, which the field-worker who first described it, Rosalind Alp, calls "foot-wear." This innovation shows that ape tools, like human hair bands and chairs, are sometimes invented as much for comfort as subsistence.[190]

And then there is the mounting evidence for self-medication. Some of it is widespread in all sort of animals, such as the eating of clay, which contains absorbent components resembling Kaopectate, a commercial drug against diarrhea and stomach upsets. But apes are also known to chew the bitter pith of certain plants and to swallow whole leaves of others, both of which are assumed to have health benefits. Michael Huffman saw chimpanzees remove the outer bark and leaves of young shoots of *Vernonia amygdalina* to extract extremely bitter juice. Nearly all these chimpanzees showed diarrhea, listlessness, and worm infections. Fecal analysis revealed a striking drop in one chimpanzee's nematode infection following bitter pith chewing, a drop not seen in chimpanzees not taking this medicine. The same plants' bark and leaves contain toxins that can kill laboratory mice, but the chimpanzees must have learned to avoid these parts and extract only the beneficial compounds. For many African ethnic groups *Vernonia* is an essential ingredient in concoctions to treat malaria, dysentery, and a number of intestinal parasites.

Perhaps the best-known discovery, by Richard Wrangham and Nishida, is the ingestion of *Aspilla* leaves by chimpanzees at Gombe and Mahale. Before consumption, these hairy, rough-surfaced leaves are carefully folded with tongue and palate so that they can be swallowed whole. Since they are not chewed, they end up in the feces undigested. Chimpanzees tend to swallow them in the morning, before foraging for food. Huffman has shown that they act as a mechanical device to expel internal parasites. Since its initial discovery, leaf swallowing has been identified in a great many chimpanzee populations across Africa, involving over thirty different plant species. Regional differences in plant selection hint at cultural transmission.

It remains unclear how chimpanzees in the forest have learned bitter-pith chewing and leaf swallowing. Experimentation with novel foods, especially those that are inedible or taste bitter, is uncommon. Perhaps the medicinal properties of certain plants were discovered by accident during a period of food scarcity. Once established, offspring watched their mothers ingest these plants and noticed the unusual way of handling them. This may have predisposed the young to try things out for themselves. For the moment, however, all of this is pure conjecture. Since self-medication seems hard to acquire without the example of others and is assumed to assist survival, it remains high on the agenda of research into chimpanzee culture.[191]

How behavior is being transmitted remains a central puzzle in cultural primatology, one that can be solved only experimentally. Fieldwork, however, yields strong hints of what to look for. For one thing, observations of primates in naturalistic settings suggest that the usual laboratory approach, with human experimenters demonstrating actions to apes, is flawed. This procedure may occasionally work, but evolution is unlikely to have equipped animals with much of a tendency to mimic strangers of their own kind, let alone those of another species. As discussed in the previous chapter, the typical context is that of BIOL (bonding- and identification-based observational learning), according to which mothers, peers, and other favorite social companions are most readily emulated. The desire to fit in sets the motivational stage, so to speak, for the actual learning process and provides its main incentive.

You Scratch My Back

The simplest kind of transmission is illustrated by the battle between milkmen and blue tits in Great Britain in the 1930s. At the time, milk was still delivered in glass bottles at the doorstep. The bottles were covered with a cardboard lid; directly underneath was the cream, which had risen to the top of the milk. Blue tits (related to chickadees) have a habit of peeling back bark to look for insect larvae; hence, removing a cardboard lid may have been a natural thing to do, and it would yield an immediate creamy reward. Once one bird had developed this habit, other birds would find bottles opened by this bird, taste the cream, and become eager to open bottles for themselves. Replacing the cardboard lid with aluminum

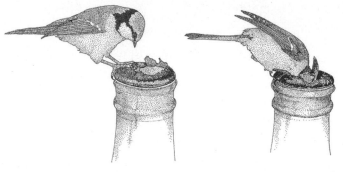

A *blue tit (chickadee) eating cream after having removed the aluminum foil seal on a milk bottle. When milk in Great Britain was still being delivered at the door, this feeding technique spread like wildfire. After encountering previously opened bottles, naïve birds may have been able to figure out the rest on their own. (Drawing by Margaret La Farge, with kind permission of the artist).*

foil caps didn't stop the birds, and the habit spread gradually from one city to all of the British isles.[192]

In my view, culture is a *social* phenomenon, however, which means that the example of others is required before I call a habit culturally learned. Milk-bottle opening by blue tits doesn't qualify, since it is unclear whether the birds actually needed to *see* other birds do it. They may have learned it as a result of environmental modification by other birds. I rate transmission in the absence of models quite differently than I rate, for example, the way the Koshima monkeys developed sweet potato washing. The fact that this habit spread first within families and among peers strongly suggests that the as-

sociation between potatoes and the ocean was learned in a social context, probably from being close to others who already
showed the habit, watching them, and tasting discarded
pieces. This satisfies my requirements for cultural learning.

To show that such examples are not limited to primates, another well-studied case is that of roof rats surviving in a pine
forest. I always see the squirrels outside my window as rats
with fluffy tails, but of course squirrels are perfectly adapted to
an arboreal life in ways that rats are not. Rats don't belong in
the trees, yet in one newly planted Israeli forest devoid of
squirrels, rats have made a living out of extracting seeds from
pine cones. A human trying to do the same with a pocket
knife quickly notes that this is a strenuous task. In the laboratory, the same rat species was deprived of food and given
cones for several weeks, but less than three percent of the
adults learned by themselves how to open them. It is not a genetic characteristic either, because young rats from mothers
that do strip cones will not learn it if they are foster-reared by
females that don't. Most likely, the young learn the technique
by being close to their mothers on foraging trips to the trees,
perhaps feeding on the same cones next to her. Even if all
they learn is that underneath the scales resides good food, the
rest of the technique can be developed by themselves. Since
they learn this in a social setting, it qualifies as a cultural tradition, and one with great survival value at that. Biologically,
these rodents are rats, but culturally they are on their way to
becoming squirrels.[193]

More complex forms of learning are those in which one individual copies the actions of another. Work on great apes, in particular, indicates that they pay attention not only to *where* but also *how* others obtain food. Duplication of another's methods, which has been demonstrated experimentally, is getting close to imitation.[194] Carel van Schaik has proposed that a precondition for such learning is tolerance. Van Schaik discovered a group of orangutans in a Sumatran peat swamp, at Suaq Balimbing, with an elaborate tool technology. Until then, the Asian red ape had never shown much of a proclivity for tool use in the wild even though its counterpart at the zoo is known as about the handiest tool manipulator of the animal world. As if to illustrate my earlier point that it is unusual for capacities seen in captivity not to be expressed also in the field, and vice versa, the Suaq Balimbing orangs poke twigs into holes to extract honey from stingless bee nests, and remove the seeds of *Neesia* fruits with short sticks that serve as both wedge and spoon. The seeds are embedded in stinging hairs and so cannot be removed by hand. The orangs modify their tools by breaking them, stripping off leaves, or chewing the end, thereby fraying them.

According to van Schaik, it is easier to concentrate on a tool task, and hence develop the most efficient technique, if one doesn't need to constantly look over one's shoulder for approaching competitors. The orangutans at Suaq Balimbing are remarkably gregarious and tolerant for their species, which further helps in the transmission of skills, as it means that others can watch at close range what the more advanced tool users are doing. This may be the secret of their special tool culture; if low-ranking individuals don't dare to come close to a dominant at work, and need to avert their gaze so as not to draw attention to themselves, it will be hard to build up a cultural repertoire.[195]

For some behavior it is unclear how it could possibly be transmitted. During my time in Japan, I heard of two striking examples, both relating to the most ubiquitous social pastime of primates: grooming. Ichirou Tanaka is conducting perhaps the finest-grained animal behavior study I have ever seen, which focuses on the exact technique by which a Japanese monkey removes louse eggs from the hair of the ones that it grooms. Louse eggs are like tiny little doughnuts stuck around the base of a hair. They need to be loosened and taken between thumb and forefinger to be drawn to the hair tip before they can be transported to the mouth of the groomer and eaten. After taking close-up videos of more than four thousand sequences of louse-egg handling, Tanaka concluded from slow-motion analysis that some monkeys are very good at it (removing the egg in a single twisting-combing movement), whereas others are, well, lousy. One might think that the difference has to do with experience, but the investigator found no relation with the groomer's age. Instead, the level of proficiency depended on the family a monkey came from. Since Tanaka doesn't believe that this sort of detailed skill is genetically inherited, it must be learned. Do monkeys really pay such close attention to each other that they notice how their relatives treat each and every single hair when they groom? It is hard to imagine, but not impossible.[196]

The second example, first recognized as a custom by Linda Marchant, concerns the chimpanzees at the Mahale Mountains. One chimpanzee will walk up to another, vigorously scratch the other's back a couple of times with his or her fingernails, then settle down to groom the other. In between the grooming, more scratching may follow. It is an extremely simple behavior, one that cannot be hard to learn for an animal that commonly scratches itself. But here's the rub: usually one scratches oneself in reaction to itching, and enjoys immediate relief as a result. Scratching someone else is really something else, for the scratcher obtains no reward.

The second puzzle is this: why is this behavior limited to Mahale? Other chimpanzee communities have been studied for decades by astute observers, yet they don't back-scratch at all. In a recent analysis, Michio Nakamura proposes two ways in which the behavior may have spread at Mahale. One is imitation of the scratching movement by a bystander. Note that the recipient cannot see the action performed on his back; hence, it is more likely that a third chimpanzee sees another do it and then copies the movement. The problem with this explanation is that imitative learning usually assumes a reward for the imitator. In the case of the social scratch, however, all benefits go to the recipient, and the imitator gets nothing out of it.

The second possibility is even more intriguing. Imagine that one chimpanzee was accidentally scratched by another, and it felt so good that he decided to offer the same pleasant experience to another chimpanzee, perhaps one that he wanted to ingratiate himself with, such as the boss. This would require perspective-taking; that is, he would have to translate a bodily experience into an action that re-creates the same experience in another individual. He would need to understand that what he felt can also be felt by another. This is a highly complex capacity, related to empathy and role reversal, for which there are indeed indications in the spontaneous helping behavior and deception of apes.[197]

This brings me to the pinnacle of social transmission short of symbolic communication: active teaching. Teaching requires the same perspective-taking mentioned above, because one needs to gauge the knowledge of the other, and judge it deficient, before one can fill in the gaps through instruction. A good teacher enters a pupil's shoes. Evidence for such a process is minimal at this point, but Christophe Boesch has seen mother chimpanzees in Taï Forest facilitate the learning of nut cracking by leaving hammers, nuts, and anvils in a ready-for-action position for their infants to use. This seems intentional, because normally adults finish all of their nuts and take the hammer with them. Boesch reported how mothers sometimes perform the cracking motions, which are very rapid, in slow motion while paying attention to the gaze direction of their offspring, as if showing them how it is done. Mothers were also seen to "correct" an unsuccessful youngster

A jackdaw explains to a class of youngsters which enemies to watch out for. Active teaching, however, is extremely rare in nature. (Drawing by Konrad Lorenz, from Er redete mit dem Vieh, den Vögeln und den Fischen © *1983 Deutscher Taschenbuch Verlag, Munich. This book, which first appeared in 1949, was translated as* King Solomon's Ring*).*

by removing a nut that it was working on from the anvil, clean the anvil, and reposition the nut, or by reorienting the hammer that the offspring was holding. Since there are so few of these remarkable observations, they may not convince everyone, yet they at least hint at a serious possibility.[198]

Perhaps the strongest evidence for teaching comes from an entirely different class of mammals. Christophe Guinet has observed killer whales, or orcas, teaching a complex hunting technique to their young. Orcas sometimes strand themselves intentionally, throwing themselves onto Argentinean beaches with breeding seals, grabbing one before they retreat to the ocean. This is a hazardous technique that, if performed incorrectly, may lead to permanent stranding and death for the predator. Adult orcas have been seen to encourage their young to strand on beaches devoid of seals. The adults push the young onto a beach, after which they help them out, in case they get stuck, by creating wash. They have also been seen to push young in the direction of prey on a beach, and if the young don't capture any, to throw their own prey at them. Since adult orcas seem to hunt better without young around, this means that they adjust their behavior in a way disadvantageous to themselves but beneficial to the young apprentices. All of this gives the impression of active instruction.

The gamut of cultural transmission in animals thus runs from picking up simple associations, such as between potatoes and water or between pine cones and seeds, to imitation of actual techniques, such as nut cracking, and perhaps even to perspective-taking and teaching of the young. It will take great effort by scientists in both the field and the laboratory to sort out which processes apply, but it is already clear that the conclusion that animal culture must rest on simple processes is premature.

Of Memes and Genes

When William McGrew published his landmark *Chimpanzee Material Culture* in 1992, a work that got the field of cultural primatology off to an excellent start, he didn't once mention "memes." Nor did Boesch and Michael Tomasello refer to memes when, in 1998, they held hands across the divide between field primatology and experimental psychology in a stimulating article in which they concluded that "in comparing chimpanzee and human cultures, we have noted many deep similarities."[199]

Nevertheless, in the minds of some people, talk of cultural transmission means talk of memes. The term compares the way cultural information is being copied with how genes send copies of themselves down the generations: memes are packages of information that spread, or don't spread, similar to the way some genes fare better at replicating themselves than others.[200] Having followed its own advice of heavy advertising, the meme concept has generated tons of references to itself. But even though the scientists just mentioned surely have heard of it, the concept is not in vogue among those who study cultural learning in animals. This is because parallels between genes and memes are about as useful as those between brains and computers. There was a time when the brain was explained as having software and hardware, as having memory banks and running programs. The urge to relate the new to the familiar—with the familiar serving as security blanket from which the new is explored—leads us sometimes into misleading comparisons. With the progress of neuroscience all of this has been forgotten; the brain is emphatically not a wet computer. Even when both handle the same task, such as playing chess, they achieve their ends in totally different ways.

Similarly, genetic evolution is a poor model for cultural change. In contrast to genetic instructions handed down from parents to children, culture is more like a set of suggestions broadcast to any recipient with an antenna to pick up the signal. New habits, such as dancing the Macarena or saying "ciao" for goodbye (except in Italy, where they have done so for a while), travel quickly in all possible directions and may reach an entire population without necessarily providing much of a fitness advantage. Most importantly, rather than coming about through random mutation, habits often result from deliberate choice: culture can be consciously created. Whereas an entire group of scientists keeps trying to shoehorn cultural transmission into the flawed metaphor of genetic evolution, students of animal behavior are wisely treating culture as a phenomenon with its own dynamics, most of which have barely been explored.

This is not to say that both forms of behavioral inheritance—the one traveling across time via genotypes, the other via phenotypes—should not or could not be conceptually linked. Ironically, the Lamarckian idea that acquired characteristics can be inherited has found its realization not in the physical characteristics he was thinking of, but in behavior. Genetic predispositions feed into culture, culture affects survival, and survival and reproduction determine which genotypes spread in the population. In other words, there exists a dauntingly complex interplay between genetic and cultural transmission. Brave and inspiring attempts at a theory of dual inheritance, or coevolution, have been made, without, however, in any way confusing the two processes.[201]

In the meantime, the chief task for ethologists and zoologists is to show that dual inheritance is not limited to our species. Apart from the already mentioned studies of cultural learning, which concentrate on the individual and its habits, a second major approach has been to compare entire groups or communities. Since this approach resembles ethnography, or cultural anthropology, the field is increasingly referred to as cultural primatology. But perhaps we should find a different term that looks beyond the primates, such as "cultural biology."[202] I have already mentioned rats and orcas, and there are countless other nonprimate examples, from the elephants in a desolate part of Namibia that have a collective, perhaps centuries-old, knowledge of a barely accessible watering hole in the mountains[203] to the food wars between bears and people in the American wilderness.

In the old days, people could protect their food by simply hanging it from a rope in a tree. Bears have caught on to this, and to more elaborate techniques, such as those involving two ropes on two trees with the food suspended between them. Bear-proof steel boxes are now recommended. Cars don't make for good boxes, as shown by bears specializing in certain car brands to get their paws on human food. They know that the windshields of certain vans can be pulled out, and that if they jump up and down on the roof of some compacts the doors pop open. These solutions seem to spread like wildfire through bear populations in a way suggestive of cultural learning.

I am convinced that the more we look for culture in animals, the more we will find it. The latest exciting finding is that of a population of capuchin monkeys—the most dexterous monkey species, with exceptionally large brains—in the Tietê Ecological Park in Brazil that cracks nuts with stones in a manner not unlike chimpanzees.[204] One day we will wonder why such an obvious idea as social transmission of knowledge and habits has taken so long to be taken seriously. Among primatologists, an obstacle has been that each field-worker has his or her own research site, often associated with a strong sense of pride, as in "My monkeys live in the harshest environment" or "My chimpanzees are smarter than yours." There is nothing wrong with this, except that the usual competitiveness of academics (because there is so little at stake, as the saying goes) hampers free information flow. Thus, some scientists have worked side by side a short distance apart for decades without ever visiting each other.

The first to set up an ethnographic program among chimpanzee researchers was McGrew, who, with his usual attention to detail, compared reports about different technologies at various African sites and began to visit colleagues to see things for himself. According to some, his conclusions—summarized in *Chimpanzee Material Culture*—should have shaken the social sciences to their roots, but they didn't. Not that his work went unnoticed; it opened the eyes of many of us to the hidden treasures lurking behind the patient chimpanzee research of the past decades, and McGrew was honored with a major prize.[205]

Things happened rapidly after that, with a first collection of papers on the topic, *Chimpanzee Cultures*, published in 1994, and finally a grand systematic survey of cultural variants in wild chimpanzees put together by Andrew Whiten and eight other authors appearing in *Nature* in 1999.[206] The latter brought cultural primatology to the attention of the public outside of our little discipline and also taught us how much there is to be gained from collaboration. Perhaps this paper will permanently alter our own culture!

The evidence in the *Nature* article, drawn from an accumulated 151 years of research at seven different field sites, included behavior patterns never published before. For example, some populations fish for ants with short sticks, eating the prey off the stick one by one, whereas at least one population has developed the more efficient technique of accumulating many ants on a long wand, after which all insects are swept with a single hand motion right into the mouth. After compilation of a first list, variants were rated on a scale from customary to absent at each site, with its ecology taken into account. For example, chimpanzees will not sleep in ground nests—as opposed to tree nests—at sites with high leopard or lion predation. Such ecologically explained differences were excluded from the list.[207] What remained were no less than thirty-nine distinct behavior patterns—far more than reported for any other animal—that vary across chimpanzee communities.

It is hard to imagine that genes would instruct apes how to fish for ants, or whether or not to make cushy seats out of vegetation. In addition, the authors found no evidence that habits vary more between than within the three existing chimpanzee subspecies. All in all, the evidence is now overwhelming that chimpanzees have a remarkable ability to invent new customs and technologies, and that they pass these on socially rather than genetically.

If animal groups vary with respect to a single behavior, such as potato washing, there is perhaps not much reason to employ a loaded term such as "culture"; "group-specific trait" or "tradition" will do. The claim has been made that chimpanzees differ from other animals in that, rather than having one or two cultural variants to go around, they have many. But I would be careful about putting the chimpanzee on a pedestal, even though there is no doubt in my mind that this species is and will remain central in any discussion of animal culture. Nevertheless, the other great apes, whales, dolphins, elephants, and many other long-lived, large-brained animals are also excellent candidates for multifaceted cultures. For example, whales sing complex songs with geographically distinct dialects. They also have elaborate hunting techniques that may be culturally learned, such as when humpbacks envelop schools of prey by clouds of bubbles. A recent study by Hal Whitehead suggests that sperm-whale mothers pass on feeding techniques and antishark defenses to their calves. He believes that some mothers are not only genetically but also culturally producing descendants that do better than others.[208]

If true, this would be dual inheritance at work! There is every reason to believe that some animals have made the step in which the struggle for life is won, at least in part, by learning from others. They don't need to find out by themselves which predators are dangerous, how to gain access to good food, which foods to avoid, how to medicate themselves, and so on. They draw on the accumulated knowledge of their family and group. In this sense, they have made the same step from nature to culture that we claim for ourselves. But it should be clear by now that I feel this is the wrong way of putting it. Thinking of nature and culture as distinct and separate domains is tricky: there's plenty of nature in culture, just as there is plenty of culture in nature.

8

Cultural Naturals

Tea and Tibetan Macaques

"But different as are the ways in which different cultures pattern the development of human beings, there are basic regularities that no known culture has yet been able to evade."

Margaret Mead, 1950

*T*ea is a necessity of life in China. Everywhere, people carry around handheld containers with floating tea leaves, occasionally replenished with hot water so as to extract a second or third serving. Tea accompanies tourists to the Great Wall, farmers to the field, employees to work, and students to class. A hotel may fail to provide shampoo or towels, but every room will have a large thermos with hot water, porcelain cups with covers, and tea bags, so that guests can brew their own drinks.

The Chinese tea habit has been exported in somewhat modified fashion to other places, such as Japan, Russia, and the British Isles. However, the habit ultimately derives from the universal need of land animals to supply their bodies with water. People require H_2O, and human culture does everything to ensure the intake: large-nippled bottles of mineral water for students on American college campuses, the absolute must of wine with Mediterranean food, and the feeling of the Dutch that even the shortest visit to someone's home is incomplete without a cup of coffee.

On a grander scale, the same applies to the thousands of cuisines and eating habits on this planet. Norbert Elias, the German sociologist for whom eating with fork and knife represented the pinnacle of refined self-control, once provoked an angry accusation of ethnocentrism from a Kenyan philosopher at an international meeting. The philosopher didn't consider himself any less civilized than Elias despite the fact that he, like millions of others in the world, ate with his hands.[209] I had to think of this bourgeois theory of civilization in the face of the versatility and elegance of chopsticks. No confusing array of utensils adapted to every course, no risk of holding the wrong tool in the wrong hand: just two little sticks for everything, from rice to an entire fish!

Whatever the variability, and despite the symbolism surrounding food, there is no denying that we all need to eat, and that what our stomachs do with what we stuff into our mouths is the same as what animal stomachs do. Culture does have the power to go against human nature—with regard to ancient China, we only have to think of foot binding, eunuchism, and ascetic monks—but some of these practices serve to suppress particular groups, while others characterize only a minority. For a culture to survive, there is a limit to the number of genitals it can remove and feet it can deform. Although the relation between culture and nature can be tense, culture mostly tries to get along with nature, like the mouse with the elephant, because there is little doubt which is the heavyweight.

Shared Humanity

In cultural studies, basic human needs are often taken for granted; they are obvious to the point of being boring. Hence the immense common ground among cultures tends to be overlooked, and differences are blown out of proportion.

I am not saying that cultural variation is a figment of the imagination or hard to see. As a trained observer and a European, I can tell a German and French family apart from a kilometer away by their demeanor and body language. But at the same time both move around as families, the universal building block of human society. I cannot agree, therefore, with cultural determinists, such as Ashley Montagu, that "man has no instincts, because everything he is and has become he has learned, acquired, from his culture, from the man-made part of the environment, from other human beings."[210] If culture really has such a profound impact, and if there really is so

little that can be attributed to human biology, why do I have
no trouble recognizing the gestures, facial expressions, and social relationships of the Chinese despite at least five thousand
years of cultural separation from the West? Why aren't the differences more radical? The Forbidden City in Beijing—four
times the size of Versailles and ten times that of Buckingham
Palace—gives a good impression of how Chinese emperors
used to rule from elaborate thrones designed to overlook the
masses gathered on immense squares, intimidating them with
their splendor. It is the same kind of formalized dominance
known of European royalty of bygone eras, or for that matter
of a chimpanzee alpha male, imperiously standing with his
hair bristling, barely paying attention to his underlings, who
crawl in the dust, approaching him with submissive pant-
grunts.

Similarly, Chinese opera has all of the same intrigues,
power plays, betrayals, jealousies, and illicit love affairs known
of Italian opera, even though the costumes, acrobatics, and
music are quite different. And in everyday Chinese life, there
is abundant familiarity in the way children whine for attention, lovers stare into each other's eyes, a good joke causes instant joy, voices are raised in anger, and longtime spouses are
in tune with each other.

In short, underneath all of those fascinating cultural differences to which we attach such great importance resides a
shared humanity that makes even the most naïve visitors feel
at home in cultures across the globe. Rather than having

landed on Mars, they are among people very much like themselves. The issue of culture always needs to be placed in this larger context of how it builds upon rather than replaces universal human tendencies. This is the oldest theme of biology—unity in diversity—and it applies to everything we consider, because the new is always constrained by the old from which it derives and with which it needs to harmonize. The same is true for animal culture, which expands behavioral possibilities and creates variety but always within a species' needs and nature.

Shared Macaqueness

Huang Shan (pronounced "fuan-sjan," meaning "yellow mountain") is *the* mountain of China, not because its 1,841 meters make it the highest peak, but because of its unsurpassed beauty. Huang Shan sticking its head out of a blanket of clouds figures in hundreds of paintings, some very famous.

Before seeing the mountains, however, my companions and I drive through gentle hills and valleys made fertile through irrigation schemes on the rivers. The rolling landscape and warm yellowish colors caused by the rows of drying rice plants remind me of the French countryside, even if the water buffaloes and large pointed straw heads of the farmers make it abundantly clear that I am someplace else.

Straddling the border between China's temperate and subtropical zones, Anhui province nowadays produces such a good rice crop that it is hard to imagine that in the 1960s people here were starving by the hundreds of thousands. It is estimated that in all of China, more than twenty million people paid with their lives for Mao's agricultural reforms. These reforms ignored the human tendency to work first of all for the family, and only secondarily for the greater good of society. Hard work comes a lot easier when it feeds hungry mouths at home than hungry mouths elsewhere. Cooperatives established under communism can then be said to have been in stark violation of human nature. The economic and ultimately political failure of this doctrine in the Soviet Union, China, and elsewhere is a large-scale illustration of the limited power of cultural revolutions. In human affairs, there are certain things that cannot be touched. This is what Edward Wilson meant when he said that human nature keeps us on a leash: we do have maneuvering room, but it is finite.

I am visiting this place not to reflect on the foibles of communism, however, nor to indulge in a sense of shared humanity, but to see a rare primate, the Tibetan macaque. Here is an opportunity to test for shared macaqueness, as I am well-acquainted with other members of the genus *Macaca*. This genus counts approximately twenty different species, ranging from the rhesus monkeys of India and Nepal to the Barbary monkeys of North Africa, to the snow monkeys of Japan. The largest macaque of all, however, dwells here in the mountain forests, *Macaca thibetana*.[211]

The species is found in a small number of scattered locations in Tibet and other parts of China. Only an estimated 450 of them live in this, the eastern part of its range. A monkey park has been set up where tourists can watch one group of around thirty monkeys attracted by regular feedings. The method of provisioning is critically important, as these monkeys have been known to kill tourists; at another site, Mount Emei, Tibetan macaque males have jumped people who crossed a narrow path along the rocks. Hitting them with their full muscle mass, they have sent several people down the precipice to their death.

Aggression against humans is mostly related to frustration when begging for food isn't rewarded. At other places, such as at Indonesian or Thai temples, I have seen long-tailed macaques threaten people, even jump on their shoulders and pull their hair, but those monkeys are so much lighter, and the footing is so much more secure, that these attacks are not lethal. The solution is to control feeding. If monkeys have no reason to search people's pockets and intimidate them, a lot of problems are avoided. Here at Huang Shan, the monkeys receive food only from park wardens.

Our little excursion is lead by Jinhua Li, a Chinese researcher who has spent the last ten years studying the monkeys. He works not just at the park, but also in the wild parts of the mountains. Following the monkeys is an almost impossible task given that they race up and down the steep rocks like mountain goats, leaving the clumsy human to carefully select his way through dense undergrowth. Li is obviously in tip-top physical shape. He talks incessantly while the rest of us struggle out of breath to the top, climbing endless flights of wooden stairs.

It is late summer, and the monkeys are not overly eager to come to the site. The forest is full of mushrooms, nuts, and berries—no need to hurry for the handouts of corn that the bipedal primates scatter around. When they finally do show up, they are truly impressive. I can see why decades ago people of this region thought they had seen the Yeti when they ran into an adult male of the species. Myths abound: they live in caves, bury their dead, are led by the oldest, and so on. Their flat, broad, bearded faces provide perhaps the most humanlike countenance I have ever seen in a monkey. They are clearly monkeys, though, not members of our immediate family. In sharp contrast to humans and apes, for example, they scratch their backs with a foot, like a dog. They also don't have the wide chest and protruding shoulders typical of the hominoids. And most tellingly, they have tails, albeit rather short ones.

The monkeys strongly remind me of stumptails, those noisy, smelly, congenial macaques I worked with for many years. Throughout my studies, I contrasted stumptails with rhesus monkeys, the quick-tempered, pugnacious New Yorkers of the primate world. Rhesus have a despotic dominance style, which means that dominants harshly punish transgressors, fiercely compete over food and water, and rarely make up after fights. It is a very one-sided relationship, with the dominant having lots of privileges, the subordinate almost none. The stumptail, in contrast, shows a high tolerance level and great concern about the state of its relationships. Stumptails reconcile with former opponents through elaborate "hold-bottom" rituals and lengthy grooming sessions. In stumptail society, inequality is mitigated, and whatever status differences are left do not seem to stand in the way of good relationships.

Tibetan macaques share a long list of characteristics with the stumptail. They walk in the same stocky, bearlike manner (stumptails are known as "bear-macaques" in Dutch and German), utter the same excited shrieks and squeals when they embrace each other, teeth-chatter in friendly situations, lock together after copulation, know sexual harassment,[212] and have brightly white babies that contrast with the dark-brown adults. The calls of both species differ from those of other macaques, yet are so similar to each other that I would be hard-pressed to tell stumptails and Tibetan macaques apart by ear. With so many behavioral similarities, it would be surprising if they weren't closely related. Probably, their temperaments and dominance styles are alike, too.

Infant Bridges

There are also differences between stumptails and Tibetan macaques, however. The most fascinating one concerns those tiny white infants. Stumptails are extremely attracted to infants in their troop, which are easy to spot due to their juvenile coat, yet they never pick them up the way Tibetan macaques do. Only in the latter species do adult males use infants as passports to get close to one another. On one occasion, an infant squealed, drawing attention to itself, when it saw the third-ranking male appear from behind a rock. The male didn't hesitate to grab the infant and turn around to take it to the second-ranking male. He approached the other male with a wide grin on his face, holding out the infant for so-called "bridging."[213] While one male held its arms and the other its feet, the infant was stretched between them at eye level, literally bridging the gap between two huge, teeth-chattering faces. The males seemed focused mainly on the infant, but also looked each other in the eyes. One male briefly sucked the infant's erect penis. Both resumed their activities afterward as if nothing had happened. The subject of all this tugging ran straight back to its mom, who had followed the scene from a safe distance.

I had to rely on Li, of course, for the males' identities and hierarchy; it takes a long time to gather such knowledge. Even the most experienced primate watcher sees only half of what is going on when he first encounters a troop. Li explained how one immigrant male, who had reached alpha status upon joining the troop, lost his rank through a social mistake: he punished a juvenile. Immediately, four high-ranking males banded together in defense to chase him off, seriously attacking him. It was as if this quick victory taught the others "Hey, we can beat this guy!" From that day on, they gave low-level repeat performances of their triumphant action and remained on top. In an attempt to gain acceptance with his victors, the former alpha male became the most frequent user of an infant passport to approach them.

The lowest-ranking male while I was there was also a former alpha who had lost one testicle in the fight that deposed him. This one-balled male hung around the periphery of the troop, unable to join the others at the provisioning site. The park wardens helped him out with separate feedings directly from their hands.

Their occasional battles notwithstanding, I was struck by the extraordinary tolerance among males. To see six big, heavily armed males (they have long, sharp canine teeth) moving about close together, almost touching each other, while collecting attractive food was a sight to behold. Yes, these males vie over females and status, but they are also perfectly capable of toning down food competition, and most of the time they get along fine. To keep the peace, they engage in excited mounts and embraces, mutual grooming, and infant-bridging. Away from the feeding site, I saw males pass each other calmly, without the nervous glances commonly seen in other species.

Apart from chimpanzees, I had never seen primate males so intensely involved with each other. In chimpanzees, too, males are at the same time rivals and friends, and I would argue the same for human males.

Social Culture

The first time I heard about Tibetan macaques was many years ago at a zoo that had acquired three "specimens" (zoo lingo for individuals). The species is virtually nonexistent in captivity, so I was anxious to see them. I was told that when the males masturbate (which they do in the absence of females), they produce an incredible amount of semen, ejaculating over great distances.

This sounds true in view of Li's observations at Huang Shan. The large testicles, the number of times a male can ejaculate in succession, the visible sperm plugs in the vaginas after copulation, the tendency of males to manually remove the plugs of other males when inspecting a female all hint at sperm competition. That is, males compete over chances to fertilize females not only through physical combat but also by dispatching as much sperm as possible toward the ovum. The injuries I saw on some males, and the tense guarding of one female by the alpha male, made it clear that aggressive competition has far from disappeared. Still, sperm competition may have partially taken its place. Whenever natural selection favors large testicles over nasty temperaments, male bonds are permitted to grow closer. The survival value of such an arrangement is as yet unclear, but one speculation is that Tibetan males collectively defend their troop. Perhaps big cats once roamed these mountains, driving the monkeys to cooperation.

Such an adaptationist explanation does not preclude a role of learning in male-male affairs. For example, it is hard to believe that infant-bridging is a simple instinct, shown by each and every Tibetan male regardless of experience. To carry an infant to another male and engage in this peculiar tête-à-tête may well require that a male has seen this behavior before, or participated in it as a youngster. Imagine a Tibetan male raised by a macaque species in which males totally ignore infants: would he still develop the bridging habit?

I raise this question because it is often assumed that behavior shown by all members of a species cannot be learned or cultural. "Species-typical" has become synonymous with "biologically determined." While not denying that biology is involved, we usually don't exclude the taming of fire from the cultural domain simply because it has been achieved by every human society. Some cultural inventions come naturally to us, such as building roofs over our heads, performing marriage rituals, or developing a classification system for close kin. There are plenty of ubiquitous cultural products.

Returning to macaques, there is one feature shared by all members of this genus: the matrilineal hierarchy. As opposed to migratory males, females remain in their natal troop all their lives, forming stable, kin-based networks between mothers, daughters, granddaughters, sisters, and nieces. It is termed a hierarchy because the rank of each female derives from this network. Since its discovery by Japanese primatologists, hundreds of studies have shown how the future rank of a young macaque female can be predicted on the basis of her mother's rank. Females with relatives in high places are born with a silver spoon in their mouth, whereas others spend all their life at the bottom, being born to a lowly matriline.

Despite its apparent rigidity, however, the system depends on learning. Early in life, the young female finds out against which opponents she can expect help from her mother and sisters. When sparring with peer A she may utter a barely audible squeak and yet receive massive support to defeat her. But against peer B she can scream her lungs out and nothing happens. Obviously, the young female will seek more and more confrontations with A—or A, who is equally smart, will learn to avoid her—whereas she will become very careful with B, who eventually will rise above her in status.

It is not that daughters of high-ranking females have blue blood; they are not inherently superior. Through adoption, for example, one can turn the infant of the lowest female into a successor to the throne, provided she is raised by the alpha female. Other experiments have manipulated the presence of family members and found that with dwindling support dominant females are unable to keep their positions.[214] In other words, the hierarchy is maintained for generation after generation through social rather than genetic transmission.

In my own work, I discovered that the same holds for social ties among female macaques. Together with my assistant, Lesleigh Luttrell, I closely followed the development of a large cohort of rhesus infants for many years, until they reached adulthood. We found that daughters copy their mothers' association preferences. Even when they have grown fully independent, and are approaching motherhood themselves, they spend much time with the daughters of their mothers' best friends.[215] Thus, the friendship between two of our females, Ropey and Bizzy, led their daughters, Robin and Bubbles, to become good friends as well.

We don't know exactly how friendships are being transmitted across the generations, but the simplest way I can think of is that when the mothers, Ropey and Bizzy, sat down to groom and relax, Robin and Bubbles would take the opportunity to play nearby. Being playmates early on, these youngsters then developed an association for the rest of their lives.

Given these processes, imagine that females in a particular group begin to strengthen ties outside their own families. Over time, such a trend will become more and more deeply embedded because their daughters will start doing the same.[216] Similarly, if mothers increasingly focus their attention on a narrow circle of kin, this will restrict the contact range of their offspring as well. If differently structured social networks are being transmitted, we may speak of different social cultures.

So, even though the matrilineal hierarchy is a universal feature of macaque society, it is not purely inborn. Based on natural abilities and tendencies, the arrangement endures because it is passed on by nongenetic means from mother to offspring: it is a "cultural natural."[217] Young monkeys are quick learners when it comes to kinship, friendship, support, and inequality, and then employ all of this knowledge when finding their place in society.

Improved Monkeys

My earlier suggestion of letting a Tibetan male grow up with a different macaque species is not as wild as it may sound. Two cross-species experiments have been conducted—one in the field, the other in the laboratory—with enlightening results.

Hans Kummer once captured female hamadryas baboons in Ethiopia only to introduce them to a wild troop of savanna baboons living in the same region. These two kinds of baboons have strikingly different societies. Kummer spent a lifetime documenting the social organization of hamadryas baboons, in which males herd a small number of females by delivering a neck-bite whenever they stray. Calling these brutal arrangements "marriages" (a shocking choice of words for a scientist who has chided others for anthropomorphism), Kummer wondered about the contribution by each sex. This is why he released hamadryas females near a troop of savanna baboons, a species that has none of this harem business. Here is what happened:

> Although no male took possession of them, each of them attached herself to a particular anubis male and groomed him. But when the female noticed that she was neither herded by the selected male nor protected from other group members by him, she behaved as freely as her savanna baboon ancestors must have done before the evolution of marriage.[218]

Hamadryas females apparently love freedom, which must mean that the social organization typical of their species is imposed upon them by the males. Conversely, when Kummer released savanna baboon females near a hamadryas troop, these females were quickly claimed by males, who trained them to stay close. The females learned that in order to avoid neck-bites, they had to act like hamadryas females, who never leave their master. They acquired this habit in a very short time, often within an hour, but never as perfectly as their hamadryas sisters. Lack of motivation to stay with a single male made them break the rules: ". . . the anubis females made life hard for their husbands. After days together they still had to be herded so often that their males finally gave up and let them go." As a result, the anubis females never achieved full integration in hamadryas society.

In another attempt to gauge the role of learning, I myself mixed rhesus and stumptail monkeys. Juveniles of the two species were placed together, day and night, for a period of five months. Having described this experiment before, here I will just mention its most conspicuous result. The rhesus monkeys, normally a quarrelsome, nonconciliatory bunch, developed peacemaking skills on a par with those of their more tolerant counterparts. Even after permanent separation from the stumptails, the rhesus showed three to four times more friendly reunions and grooming after fights than is typical of their species.[219]

Our new and improved rhesus monkeys demonstrate the power of social learning. True, all macaques know reconciliation, but at the same time the behavior is an acquired social skill. How could it be otherwise with animals that take four to five years to mature? A slow development leaves ample room for the environment to exert its influence. Similarly, even though all hamadryas baboon troops show the typical harem structure, it is the product of learning. Females of this species respond to social rules enforced by the much larger males.

Primates thus form cultures in which social arrangements are taught and acquired. Most of the time we barely notice the degree to which this happens, except when we mix species, switch infants, or create new environments. Sometimes, however, cultural developments result in dramatic differences between groups. For example, the Wisconsin Primate Center used to keep two large rhesus groups in identical enclosures. When I worked there, I could see, side by side, one group in which males fiercely competed every mating season and another group in which they did not. In the first group, the alpha male would not allow any other male in his sight to mount a female. This didn't stop the other males from mating; they seized every opportunity when the old leader had his back turned. Other high-ranking males were equally intolerant, however, prohibiting matings by males below them.

In the other group, five adult males might each be consorting with a female in plain view of the others without any interference. Of course, the alpha male had first pick, but he tolerated the others' sexual activity. Both groups had many more females than males, so the difference cannot be attributed to the relative availability of females.

I studied these groups for a decade, and the contrasting male relations persisted despite changes in the hierarchy. Probably, young males growing up in each group received quite contrary lessons about how to act during the mating season.

The Welcome Tree

Cultural naturals defy the traditional dualism between culture and nature. They are not cultural products, in the strict sense; nor do they conflict with biology. Thus, the matrilineal hierarchy of macaques arises automatically out of natural tendencies, such as the inclination to support kin, but these tendencies need to be supplemented with learning; otherwise a stable structure can never emerge. Similarly, the human incest taboo, long held up as a prime example of our ability to subjugate nature, is now considered a cultural fortification of a natural tendency. Like many animals, people tend to avoid sexual relations with individuals with whom they grew up. The universality of the incest taboo reflects a happy marriage between nature and culture.[220]

Cultural naturals are not to be taken lightly. Although the role of learning may create the illusion of flexibility, it is impossible for our species to get around certain cultural institutions. In the same way that communism floundered because it went against human economic nature, rules against stable family arrangements tend to backfire. The hippie communities of the sixties, based on a denial of sexual jealousy, evidently didn't last. And attempts to break the oldest mammalian union, the mother-child bond, have been equally unsuccessful. The kibbutzim in Israel have largely abandoned their communal child-rearing practices in recognition that parents and children belong together. And after a flirtation in U.S. maternity wards with purely physical medical care and incubators—resulting in the swift removal of newborns from their mothers lest they might catch infections—there is now a growing awareness of the importance of early body contact and attachment: "After a period of about ninety years during which that bond was ignored, dismissed, shattered, and fully re-examined 'scientifically,' Western culture has now returned to accepting that babies and mothers are a natural pair."[221]

I had to think of the importance of family arrangements, including the human pair-bond, when visiting Huang Shan. Somewhere on the tallest mountain there is an ancient, knotty tree, a member of *Pinus hwangshanensis* (the pine tree of this region), which is so familiar in this nation of a billion people that when foreign dignitaries visit they are often photographed in front of a large watercolor of the tree.

The image behind them does not show the tree species in general, but this particular tree, known as the "Welcome Tree." The poor old pine is propped up by a metal frame so that it can serve for another century. Approach to the tree is obstructed by heavy chains, weighed down by thousands of padlocks. No lock on the chain is single. All of them hang in pairs, each pair permanently hooked together. Chinese lovers and newlyweds visit this site, link their padlocks together, then throw the keys down the mountain.

Their gesture may not be universal, but it is universally understood. The large number of customs that people don't need to explain to members of other cultures indicates how much our societies are constructed around a shared humanity.

Section 3

Human Nature

The Way We See Ourselves

Every language is full of expressions linking human and animal behavior. In English we have idioms such as puppy love, the lion's share, horse sense, getting someone's dander up, a dog-eat-dog world, and taking someone under one's wing. Similarly, many of science's best ideas about human behavior derive from the study of animals. Either consciously or unconsciously we see ourselves through the prism of the larger nature of which we are part. But we also abuse nature by projecting our views onto it, after which we extract them again, circularly proving whatever view we hold. If studies in animal behavior teach us anything, it is that there are no simple lessons. Behavior that may look unselfish on one level may be self-serving at another. Given this multilayered reality, we should be particularly wary of catchy metaphors. Metaphors have the power to either instill or undermine confidence in our species' moral capacities, and they may do so rightly or wrongly. Debate about these issues is so essential to our self-image that it goes back several millennia in both the East and the West.

9

Apes with Self-Esteem

Abraham Maslow and the Taboo on Power

"I put for a generall inclination of all mankind, a perpetuall and restlesse desire of Power after power, that ceaseth onely in Death."

Thomas Hobbes, 1651

*F*ew people realize that Abraham Maslow, one of the first modern psychologists to explore human ambition, was greatly inspired by monkeys. He was struck by the cocky, confident air of the top monkey, and the slinking cowardice, as he called it, of individuals near the bottom of the hierarchy. Maslow also saw that high status pays off in terms of access to resources. In 1936, he postulated a drive for dominance, while rejecting the idea that those who are dominated are truly "submissive." In his mind, the latter term implies that subordinates give up any hope of besting their superiors, which he felt they never do.[222]

Turning his attention to human behavior, Maslow observed in some people the same self-confident attitude that he had seen in his monkeys. After proposing some now-forgotten labels for this attitude, such as "dominance-feeling" and "ego-level," in 1940 he hit on the concept of "self-esteem." This blend of self-evaluation and self-love struck a chord in American culture, and the term remains immensely popular. Feeling good about oneself has become a goal in and of itself, sometimes quite disconnected from actual ability or merit.

But biologists seeking an explanation for the pervasiveness of social inequality—found in chickens, wolves, horses, primates, and a host of other animals that live in groups—obviously want to know more than just how it *feels* to be on top. Self-esteem as a goal has absolutely no meaning in a harsh world of survival. What is in it for those who achieve it? If high-status males are sexually more successful than other males, either because they can lay claim to more females or because they are more attractive, they are at an advantage in the evolutionary race. If they are able to beget more offspring, genetic traits that helped them get to this point are passed on to the next generation. Animals do not think in terms of progeny, but they do act in ways that help spread their genes. Some people would agree with Thomas Hobbes that the human male has inherited the same tendency and follows it just as blindly. In modern society, we have no lack of political scandals to remind us of the age-old connection between power and sex.

For females, the number of males they mate with has little effect on reproductive success, so sex and power are separate issues. Having high rank does pay off for females in terms of food and protection of their offspring, but not in terms of attractiveness. In our own species, it appears that male sex appeal is greatly enhanced by a high position, whereas female sex appeal is not. A prominent French politician once compared power to pastry, saying that she liked it while knowing it wasn't good for her.

Maslow's interest in ambition and its potential benefits seems just as relevant today—to all primates. Indeed, I have always felt a certain bond with Maslow, partly because I share his belief in the existence of a drive for dominance, and partly because I have worked for years at the little Vilas Park Zoo, in Madison, Wisconsin, where he conducted his primatological studies in the 1930s as the very first graduate student of Harry Harlow. When people talk about self-esteem, therefore, the first image that flashes through my mind is the dignified self-assurance of Mr. Spickles, the old boss of the troop of rhesus monkeys that I knew so well. Spickles was a fully self-actualized kind of guy, never the slightest bit intimidated by even the most vigorous younger males. He had seen all of them grow up and had played with them, but he had also punished them for youthful transgressions.

Perhaps as a result, these males were psychologically inhibited in Spickles' presence, even though he had lost his physical vigor along with most of his teeth. In the wild, an old leader has to deal with strange males entering his troop, who of course have fewer scruples about challenging him. But even then, it is not always just a matter of which male is the strongest or fastest, because the collective support of the females may keep a male in the saddle well beyond his prime. They often prefer a familiar, predictable leader over a younger, aggressive upstart.[223]

The possibility of an individual being dominant yet dependent on others, as a male sometimes depends on a coalition of females, does not seem to have crossed Maslow's mind. He thought mainly in terms of individual differences and personality types. Because he saw dominance as a sign of inherent, biological superiority, he felt that in a good society the elite should be given the opportunity to realize their potential. They should be protected against the malice of the nongifted, who inevitably have trouble accepting their miserable positions. He thus forgot that dominance is a *social* phenomenon that resides in relationships, not individuals. Alone on an island, the biggest boss is no boss at all. Individual abilities do play a role in achieving high status, but the abilities involved are often distinctly social, such as diplomacy and a talent for building lasting partnerships.

Dallas Cullen, a business professor at the University of Alberta, recently examined the history of Maslow's ideas,

which remain extremely influential in the management community. She concluded that Maslow did not look carefully
enough at the role of social context in the monkeys that he
studied:

> He overestimated the autonomy of the dominant individ-
> ual, and instead saw this individual as able to function in-
> dependently and separately from others in the social set-
> ting. Concurrently, he underestimated the extent to which
> the dominant individual needed to pay attention to social
> links with others and use interpersonal skills in order to de-
> velop and maintain those links.[224]

For better or worse, our self-perception is never animal-free.
Sometimes the animal derivations of an idea are hard to trace,
as with Maslow's hierarchy of needs, yet most of the time it is
rather obvious how much our understanding of learning and
conditioning, parental care, sex and hormones, aggression,
and so on, are steeped in the realization that we are animals
with animal tendencies. Even if authors stress how different
our species is, such as when they say that culture is what
makes us human, they are still using animals as referents.
There is no escape: human behavior is always placed in this
larger context of other behaving organisms.

Inequality is a case in point. Ever since the discovery of the
pecking order, we realize that we are barely unusual as a hierarchical species.[225] At the same time we may be uncomfortable

with the comparison, as it lends a ring of inevitability to tendencies that we dislike in ourselves. I speak here as a baby boomer, who questioned authority at every turn. We challenged the bureaucrats and intellectual mandarins of the academy, expressing our defiance with long hair and extravagant clothes. It is interesting to reflect on this now that my own generation has come to power, showing remarkably little aversion to wielding it. It must be that, viewed from above, the system looks different, and far more acceptable!

Arnhem Revisited

The social side of the drive for dominance, called politics, depends on jockeying for position by seeking support, respect, and popularity. Chimpanzee males spend so much of their time and energy struggling for dominance that they come across as power-hungry Machiavellians. After six years of chronicling the social dramas in the world's largest chimpanzee colony, at the Arnhem Zoo in the Netherlands, I summarized their schmoozing and scheming in my 1982 book *Chimpanzee Politics*. Since then, field studies on chimpanzees have confirmed the political nature of these apes.[226]

Of the numerous primates that I have come to know in my life, the ones encountered while I was still a student still occupy the softest spots in my memory banks. Even though I now live across the Atlantic, I visit my native country regularly and stop by the zoo whenever I can. I am still recognized by the older generation. Mama, who must be close to fifty, unerringly picks out my face from amongst hundreds of visitors, and moves her arthritic bones to the moat to greet me with pant-grunts. But Gorilla is perhaps the happiest of all. Ever since I taught her how to bottle-feed an adoptive daughter, twenty years ago, we have enjoyed a close bond. Without this intervention she might never have raised any young. Both Mama and Gorilla now have grandchildren, and this younger generation looks at me as a stranger who, astonishingly, acts as if he belongs. The colony includes approximately thirty apes, and remains the largest and most successful of its kind.

Then there are the males, the big guys, who strut around with their hair standing up, occasionally castigating one of their underlings. They eerily remind me in manner and voice of the males I knew, but that is because they are their sons. In captivity, male chimpanzees live ten years less than females on average, and the difference may even be greater in the wild. Male lives are full of stress and tension, not to mention the physical risks when they fight one another or make spectacular escapes, jumping from tree to tree. None of my original male players is still around.

The only way to describe the events in Arnhem, I felt, was to bring out the chimpanzee personalities, and pay attention to actual events, as an unfolding soap opera, rather than abstractions and theories. In science, however, items that cannot be statistically evaluated and graphed run the risk of being tossed aside as mere anecdotes. True, it is hard to generalize from single events, but does this justify the contempt in which they are being held? Consider a human example: Bob Woodward and Carl Bernstein describe, in *The Final Days*, President Nixon's reaction to his loss of power:

> Between sobs, Nixon was plaintive. . . . How had a simple burglary . . . done all this? . . . Nixon got down on his knees. . . . [He] leaned over and struck his fist on the carpet, crying, "What have I done? What has happened?"[227]

Nixon is the only person in U.S. history to resign the Presidency, so this account can't be much more than an anecdote. Yet does that fact make the story less significant? I must admit to a great weakness for rare and peculiar events. The chimpanzees that I studied had tantrums just like Nixon's (minus the words) under similarly stressful conditions. When Yeroen, the oldest male at Arnhem, was in danger of losing his top rank to another male, he would in the middle of a confrontation suddenly drop out of a tree like a rotten apple and writhe and squirm on the ground, screaming pitifully, waiting to be comforted by the rest of the group.

The expression "being weaned from power" is particularly apt, because Yeroen's relapse into childlike behavior was the same as that of a juvenile being weaned from its mother's milk. Despite its noisy protests, the juvenile keeps an eye on mom for any signs that she might change her mind. Similarly, Yeroen always noted who approached him during his tantrums. If the group around him was big and powerful enough, and especially if it included Mama, he would gain instant courage. With his supporters in tow, he would rekindle the confrontation that he had been losing. Clearly, Yeroen's tantrums were yet another example of deft social manipulation.

Nikkie's Ghost

Since those days, many interesting developments have occurred in Arnhem. Some of these were gruesome, such as the killing and castration of one of our males by two rivals. This incident fundamentally changed my perception of the need for compromise among chimpanzees: its absence can have horrible consequences. And it is not only in captivity that these things happen. Twelve years later, a male in Gombe National Park received very much the same treatment from his own group mates. He would probably have succumbed to the ensuing infections had it not been for veterinary intervention.[228]

After I ended my work at Arnhem, Otto Adang, the researcher who succeeded me, found even more evidence for power politics among the chimpanzees there, including shifting alliances and the importance of social support for the contenders. Two males—Yeroen, the former leader, and Dandy— had banded together to oust Nikkie as the alpha male. Their alliance drove Nikkie to a desperate attempt to escape. Unfortunately, he drowned trying to make it across the moat that surrounds the chimpanzees' island at the zoo. The newspapers dubbed his death a suicide, but it seemed more likely a panic attack with a fatal outcome.

With Nikkie's death, the closeness between Yeroen and Dandy evaporated. Rivalry, predictably, took its place.[229] About a year later, in 1985, Adang decided to show the chimpanzees a movie, *The Family of Chimps*, a documentary filmed at Arnhem when Nikkie was still alive and in charge. With the apes ensconced in their winter hall, Adang wanted to gauge their responses to two-dimensional images, which he projected onto a wall. It remains unclear whether the apes recognized the actors, until a life-sized Nikkie appeared. At that point Dandy immediately ran screaming to Yeroen, jumping literally in the old male's lap! Yeroen, too, had an uncertain grin on his face. Nikkie's mysterious resurrection had temporarily restored their old pact.

Opportunism is a major part of chimpanzee politics, and most of us would not hesitate to use the same term for its human counterpart. The booing and shouting between ideological factions, and the occasional hair-pulling and throwing in the parliaments of emerging democracies, hint at a history of our political systems that is incompletely captured in the lofty terms that historians and political scientists reserve for it.

If we follow Harold Laswell's classical definition of politics as a social process determining "who gets what, when, and how," there can be little doubt that chimpanzees engage in it. Since in both humans and their closest relatives the process involves bluff, alliances, and divide-and-rule tactics, a common terminology is warranted. The title of my book drove this point home. Not all scholars were comfortable with this, of course. But Newt Gingrich, then the U.S. Speaker of the House, recognized the animal parallels when he put the Arnhem saga on the recommended reading list for freshmen representatives, in 1994. Reading about chimpanzee power plays may help politicians recognize elementary political strategies that they themselves probably apply unconsciously. Moreover, they will learn that despite the constant jockeying for position there is a certain internal logic, even morality, to the emerging social system. Success is not just a matter of wiping out the opposition. In the wild, male chimpanzees depend on one another for hunting and territorial defense: compromise and reconciliation are as much part of political skills as fighting ability.

Power by Another Name

What has always fascinated me about chimpanzee society—and, by extension, human society as well—is the impossibility of stability. Primatologists tend to speak of "social organization" as if social life revolves around a fixed structure, the backbone of the group. But the structure is more like a river, always there but never the same. Whenever one believes that events have settled down and is prepared to declare the end of history, or some such silly concept, one usually can detect undercurrents of change. A young male is growing up and beginning to make waves. An old chap is starting to tire of lengthy charging displays, and his rivals are taking notice. Some females are building large families, which grow into influential cliques.

Each time I visit Arnhem, the caretakers and students need to fill me in, as though I have missed several episodes of *Big Brother*, the popular real-life television drama: "Chimp A now is leaning toward chimp B, who is reluctant to grab power because he was beaten up by chimp C. But once his wounds have healed, the females will help him get even, because they like him a lot more than they like C." There is constant movement, along with a perpetual laying of little bricks of power on top of each other until the existing order has become ripe for yet another challenge.

In humans, we cloak these tendencies in all sorts of euphemisms. Studying status inequalities in a society with a strong egalitarian bent, Mauk Mulder, a Dutch social psychologist, spoke of the taboo surrounding the term "power." Corporate managers would tell the investigators that they enjoyed responsibility, prestige, and authority, but no, power was really not what they were after. They did recognize the lust for power in *others*, but they themselves only wanted to put their personal capacities as adviser and leader at the service of their organization.[230]

Similarly, no political candidate will publicly admit to getting a kick out of power. Politicians want us to believe that it is for us and for their nation that they sacrifice their private lives. And even when anthropologists draw direct parallels between human and primate dominance relations, they avoid the P-word, such as when Jerome Barkow concluded in 1975 that

> [w]ith the development of a sense of self, our ancestors' tendency to seek high social rank would have been transformed. Having a self means that self-evaluation is possible. The social dominance imperative would have taken the form of an imperative to evaluate the self as higher in rank than others. To evaluate the self as higher than others is to maintain self-esteem.[231]

Hence, early humans, like the Dutch managers, were seeking, not power over others, but rather some sort of pride in themselves. This is a strange thought, because what good does pride do? Only if there are serious payoffs associated with it can we explain the amount of energy put into its pursuit. Psychological explanations in terms of self-esteem all falter at this point: self-esteem has no value unless it varies in proportion to the esteem received from others and the privileges derived from it. In other words, self-esteem cannot be that important unless it is socially constructed.

This brings us back to Abraham Maslow and his ideas about self-actualization. Had he paid attention to all of the interdependencies among the primates that he studied, he might have developed a different view. Evaluation of the self does not take place in a vacuum; it requires continuous interaction with others. Absent the esteem of others, self-esteem is hollow.

By translating what is essentially a social process into an internal experience based on personal ability, Maslow deftly upheld the taboo on power. This taboo concerns how power affects others, how it constrains and directs the behavior of others. No one will deny that some people exert more influence than others, but people do not wish to hear anyone express an actual *desire* to do so. Such a desire would fly in the face of democratic ideals.

This must be the reason why social psychology textbooks hardly ever refer to concepts such as "dominance" or "power." It is almost as if the decision has been made that by not mentioning one of the strongest driving forces of the human species, it will go away. Everywhere around us we see status hierarchies—at school, in church, in the military, in business— but as an area of research it is barely developed. And so we continue to juggle the hot potato of power. It is the sort of collective lie that Niccolò Machiavelli broke with—an audacity that failed to do his reputation much good.

I am just grateful that I study social inequality in creatures who express their needs and wants blatantly, without coverups. Language is a fine human attribute, but it distracts almost as often as it informs. When watching political leaders on television, especially when they are under pressure or in debate, I sometimes mute the sound so as to focus better on the eye contact, body postures, gestures, and so on. I see the way they grow in size when they have dealt a verbal blow, or how they shut off unpleasant information by closing their eyes a fraction of a second too long. What is going on is immediately familiar

Watching adults is a favorite activity of young chimpanzees. This way, they pick up knowledge about sources of food and feeding techniques, such as this female's way of picking grubs out of rotten wood. (Arnhem Zoo, photo by author). Below, a status ritual between two adult male chimpanzees. The male on the left has his hair up and walks bipedally, brandishing a piece of wood in his right hand, while the male on the right avoids with pant-grunts, which is this species' recognition of another's high rank. (Arnhem Zoo, photo by author).

Chimpanzees and bonobos are readily distinguishable by both sound (the bonobo's voice is much higher pitched) and sight. Here, an adolescent male chimpanzee, above, and a same-aged male bonobo, below. Note the elegant build of the bonobo, with fine facial features, narrow shoulders, small ears, and reduced eyebrow ridges. (Yerkes Primate Center and San Diego Zoo, photos by author).

An adolescent male bonobo grooms an adult female. Hand-clap grooming is a unique tradition among the San Diego bonobos, in which the grooming individual interrupts its activities to clap its hands (or feet) together, making for an audible performance. (San Diego Zoo, photos by author).

A group of chimpanzees has gathered at the cracking site in the forest of Bossou, Guinea. Two adult females use hammer and anvil stones to open palm nuts, while an infant follows its mother's movements. The infant will occasionally get a kernel, and will soon reach the age to begin experimenting with stones and nuts, years before it will have the force and coordination to actually crack anything. (Photo by Tetsuro Matsuzawa, with kind permission).

A bonobo male plays blind-man's bluff, a game developed by his older female playmates. The females would completely blind themselves by poking both thumbs in their eyes, whereas this male occasionally sneaks a view of the climbing frame from underneath his arm. (San Diego Zoo, photo by author).

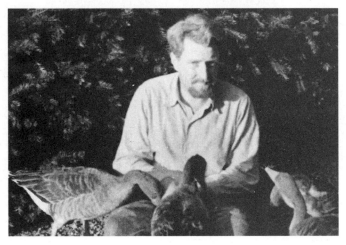

Top, Kinji Imanishi, the father of Japanese primatology, photographed in 1958 at the age of 56. (Photo by Jun'ichiro Itani, with kind permission of the Imanishi family). Bottom, Konrad Lorenz, the Austrian Nobel-Prize winner famous for his work on imprinting, surrounded by his favorite birds, in 1936, at the age of 33. (Photo by Alfred Seitz, with kind permission of the Lorenz family).

Animals sometimes make great sacrifices by simply following their maternal instincts, which evolved for the sole purpose of raising their own young. Here a mother dog which raised three tiger cubs at a zoo, steps without fear over the head of one, who could easily devour her. (Lop Buri, Thailand, photo by author).

With the Welcome Tree, a national symbol of hospitality, in the background, a chain is weighed down by locks, all hanging in pairs. The locks belong to Chinese lovers and newlyweds. After having placed them on the chain, they throw the keys down the mountain. (Huang Shan, China, photo by author).

to someone who has seen chimpanzees strive for dominance. Instead of trying to change their self-perception—which they could do equally well at home or with the help of a therapist—these politicians are working hard on gaining the upper hand in the eyes of the millions watching them. They are playing for an audience, promising, explaining, pleading, and lying. They know better than anyone else that power is not personal but interpersonal.[232]

Had Maslow looked beyond the contrasting demeanors of his dominant and subordinate monkeys, and focused on the social matrix in which both are firmly embedded, we might now have a more socially grounded theory of motivation, one more in tune with the human soap operas that we call our lives. The dominance drive is unabashedly aimed at subjugating others and exerting one's will. It is not a nice, cozy, feel-good drive, but the product of a long heritage of resource competition in which some parties fare better than others. Whereas the pursuit of power can be fully conscious, the one-upmanship and manipulations we see around us are often no more conscious than those among chimpanzees. Put a bunch of left-leaning professors with an egalitarian ethos together in a room—a situation not unfamiliar to me—and one will still see a power structure emerging. It is automatic.[233]

No Simple Lessons

The moral of the story is that animals can be used to confirm ideas that apply neither to them nor to the species for which the ideas are ultimately intended. We tend to draw on nature to make our case no matter what the case may be.

In a different, more recent example, Judith Harris, in *The Nurture Assumption* (1998), uses animal examples, specifically chimpanzees, to show how development is determined as much by nature as nurture, and as much by the peer group as the immediate family. In doing so she challenges a whole generation of scholars who have claimed that child development is an entirely cultural, educational affair, steered mostly by the parents. Who is right or wrong is not the issue here. The point is that the animal evidence is adjusted by both sides to fit their respective views.

Those who believe that human childrearing is uniquely cultural, and that other species merely sharpen their instincts, forget about the lengthy development of many animals. Given that chimpanzees are considered fully adult only at the age of sixteen, their youth is one long learning experience. And if there is one species in which parental influences should not be underestimated, it is also the chimpanzee. Youngsters travel with their mother and other dependent siblings for the first eight years of their life, with only intermittent contact with the rest of the community. The mother's behavior and temperament can therefore be expected to have a tremendous impact, much greater than that of the peer group. In short, chimpanzee development seems to support neither the we-are-uniquely-cultural nor the parents-don't-matter school of thought.

There are plenty of assumptions about animal behavior in the social sciences; some right, but many wrong. This situation will not change so long as behavioral research on people is largely divorced from that on other animals. Those who like to reach into the grab-bag of nature whenever convenient need to understand that there are no simple lessons. Just as our own kind is both noble and evil, both selfish and altruistic, both slave of and master over its instincts, the animal world shows the same contradictions. Scholars will continue to compare and contrast human and animal behavior, but hopefully less in order to support preconceived notions than to uncover the far more complex, diverse, and multilayered reality that nature holds in store.

10

Survival of the Kindest

Of Selfish Genes and Unselfish Dogs

"How selfish soever man may be supposed, there are evidently some principles in his nature, which interest him in the fortune of others, and render their happiness necessary to him, though he derives nothing from it, except the pleasure of seeing it."

Adam Smith, 1759

"Altruism may arise in the chimpanzee, in some modest degree, where there has been no training in generosity. On any reasonable view, this requires reinterpretation of the traditional hedonistic, law-of-effect view of human nature and human motivation."

Donald Hebb, 1971

*T*he most absurd animal exhibit I have ever seen was at a small zoo in Lop Buri, Thailand. Two medium-sized dogs shared a cage with three full-grown tigers. While the tigers cooled their bodies in dirty water, the dogs moved around, hopping unconcernedly over the huge striped heads that rested on the concrete rim of the pool. The dogs were walking snacks, but the tigers evidently failed to perceive them as such.

I learned that one of the dogs had raised the tiger cubs along with her own puppy, and that the whole family had happily stayed together. The mother was said to be top dog over everyone else.

The tigers were no pushovers, though. They silently stalked the three-year old son of my hosts when he strolled by the cage, their yellow eyes glued to the boy, ready to pounce if some miracle removed the bars holding them back. In the forest, a member of the same species once roared at the boy's father, a tall German primatologist, making his blood curdle, and permanently changing his perspective on the risk factors of his job.

A couple of meters from this exhibit stood a statue depicting combat between a tiger and an eagle, both of them larger than life. The eagle seemed to be trying to scratch out the tiger's eyes with its talons, an implausible encounter because the two animals normally don't get in each other's way. But it was a dramatic rendition of the ubiquitous struggle for existence, the cutthroat competition between organisms over limited resources, or, as Tennyson immortalized it, "nature, red in tooth and claw."

Both the statue and the cage with tigers and dogs presented artificial situations, but with conflicting messages. While the animals demonstrated how well teeth and claws can be held under control, the statue arrogantly declared: "Who cares what you actually see in nature? This is how it works!" Unintentionally, the zoo thus offered grounds for reflection on observed versus theorized nature.

The incredible sacrifice of the mother dog in rearing three tigers falls under the biological definition of altruism—that is, she incurred a serious cost for the benefit of others. She didn't do it for herself, her family, or even her species, so why did she do it? What energy she must have put into raising three giant animals so totally unlike herself! The difference in size was every bit as large as that between, say, a tiny hedge sparrow and the enormous cuckoo nestling she is raising. But the hedge sparrow had been tricked by an egg similar to her own, whereas it is hard to imagine that a dog is unable to tell a tiger cub from a puppy by sight, let alone smell.

Biologists often explain altruism by so-called kin selection. Kindness towards one's kin is viewed as a genetic investment, a way of spreading genes similar to one's own. Assisting kin thus comes close to helping oneself. Sacrifices on behalf of kin are pervasive, from honey bees that die for their colony by stinging intruders to birds—such as scrub jays—that help their parents raise a nest full of young. Humans show the same bias toward kin, giving rise to expressions such as "Blood is thicker than water." No wonder awards for heroism are rarely bestowed on those who have saved members of their own family.

The bitch of our story qualifies as a heroine, though, since she gave tender loving care and nourishing milk to individuals that could not possibly be her relatives. Kin selection, therefore, cannot explain her behavior. The alternative hypothesis is the "You scratch my back, I'll scratch yours" argument, where the help is directed to someone willing to repay the service. In my own work, I have tested this idea by recording grooming sessions among chimpanzees at the Yerkes Primate Center's Field Station, near Atlanta, after which I watched food sharing among the same apes. I found that if chimpanzee A had groomed B in the morning, A's chances of getting food from B in the afternoon were greatly improved. All parties stand to gain in such an economy of exchange.

Could this account for the dog's behavior? It might be argued that the cats repaid her by not devouring her, but such altruism-by-omission is a bit of a stretch. It certainly doesn't explain the mother's generosity. Had she simply rejected the cubs, she would not have had to contend with them as dangerous adults to begin with. Clearly, she got little or nothing out of the whole deal.

Does this mean that the evolutionary paradigm is fundamentally flawed? The answer depends on how broad or narrow a vision of evolution one embraces. The above theories explain cooperation reasonably well, but they do not apply—and do not need to apply—to each and every single instance.

The beauty of unnatural arrangements, such as placing tiger cubs on a dog's nipples, is that they expose the disjunction between motive and function. The original *function* of maternal care is obviously to raise one's own offspring, but the *motivation* to provide such care reaches beyond that function. The motivation has become strong and flexible enough to reach out to other young, even those of other species, regardless of what is in it for the mother. Motives often acquire lives of their own. As a result, they do not always neatly fit biology's dominant metaphors, which emphasize ruthless competition.

The Spider and the Fly

Anyone who has seen the film *Il Postino* (The Postman) realizes the extraordinary lure of the metaphor. The apprentice poet of the movie learns to offer a fresh look at the world through carefully selected analogies. Shy at first, he soon relishes the poet's proverbial "license" to transform reality, which helps him greatly in wooing the opposite sex.

People are animists by nature, always interpreting reality in their own image. It starts early when children freely ascribe inner lives to clouds, trees, dolls, and other objects. This tendency is commercially exploited with pet rocks, chia pets, and Tamagotchi, which show remarkably little resemblance to the usual recipients of human love.[234] The phenomenon is not even limited to our species; chimpanzees, too, care for imaginary young. Richard Wrangham observed a six-year-old juvenile, Kakama, carry and cradle a small wooden log as if it were a newborn. Kakama did so for hours on end, one time even building a nest in a tree and putting the log into it on its own. Kakama's mother was pregnant at the time. The field-worker notes: "My intuition suggested a possibility that I was reluctant, as a professional skeptical scientist, to accept on the basis of a single observation: that I had just watched a young male chimpanzee invent and then play with a doll in possible anticipation of his mother giving birth."[235]

Scientists are not immune to the urge to project needs and desires onto inanimate objects. Unfortunately for us, however, we lack the license of the poet and the innocence of the child. Metaphors are used in science to great effect and advantage, but also at great peril. Taken literally, they often obscure the truth. This lot befell the well-known "struggle for existence" view of the natural world. It kept generations of biologists from recognizing the shared interests among individuals and species even though Charles Darwin—always wiser than his followers—had warned in *The Origin of Species*: "I use the term Struggle for Existence in a large and metaphorical sense including dependence of one being on another."[236]

In chemistry and physics, metaphors are common, as when we say that elements are "attracted" to each other (not to mention that they "like" each other), or when we use concepts such as "force" and "resistance." Anthropomorphic interpretations are attempts to make sense of the world around us. In modern biology, this has led to the characterization of genes as "selfish" and of organisms as "adapting" to their environment. Genes are said to be our rulers, and to strive for their own replication. But really, all that is going on is that genes, a mere batch of DNA molecules, replicate at different rates depending on the success of the traits that they produce. Rather than doing the selecting themselves, genes are *being* selected. Adaptation, too, is a blind and passive process resulting from the elimination of less successful forms. All of this is known to every biologist, but we are unable to resist infusing evolution with direction and intent.

It is only a small step from calling genes selfish to slapping the same label onto the carriers of those genes: plants, animals, and people. Thus, according to George Williams, one of the world's leading evolutionary biologists, "natural selection maximizes short-sighted selfishness."[237] He thus extends the utilitarian language of his discipline to the domain of motivation. This is a slippery extrapolation, because the selfishness of genes is entirely metaphorical—genes have no self, hence cannot possibly be selfish—whereas animals and people do qualify for the literal application.

Thus, the concept of "selfishness" has been plucked from the English language, robbed of its vernacular meaning, and applied outside of the psychological domain where it used to belong. It is now often used as if it were a synonym for "self-serving," which of course it is not. Selfishness implies the *intention* to serve oneself, hence knowledge of what one stands to gain. Without such knowledge, selfishness is a much more problematic concept than many evolutionary thinkers realize. A vine may serve its own interests by overgrowing and suffocating a tree, but since plants lack intentions and knowledge they cannot possibly be selfish except in a rather meaningless sense.

The question then becomes whether animals and people possess the knowledge to act selfishly. In nature, the future is mostly hidden behind a veil of ignorance. The spider builds her web in order to catch flies and the squirrel hides nuts to get through the winter, but it is unlikely that spiders and squirrels do so knowingly. This would require previous experience, whereas even the youngest, most naïve spiders and squirrels weave webs and store nuts. They have no clue how useful their actions will turn out to be. Both species would have become extinct long ago if it were otherwise. And these are only the simplest examples I can think of. Many behavioral functions are much harder to recognize. The stallion fights at great risk against other stallions so as to claim a harem of mares and sire offspring with them, but it would be ridiculous to suggest that the stallion himself knows how a victory might affect his reproductive chances. For this, he would need to know the relation between sex and procreation, an understanding yet to be demonstrated in any nonhuman animal.

Even human behavior doesn't necessarily depend on awareness of its results. The healthy appetites of children and pregnant women, for instance, serve their need for growth. It would be a mistake, however, to assume that these individuals eat out of a desire to grow: hunger does the trick. Motivations follow their own rules, fulfill their own goals, and require their own set of explanations.

Instead of the piecemeal evolution of individual acts—such as bite, scratch, flee, lick, or nurse—natural selection has produced entire psychologies that orchestrate a species' whole repertoire of behavior. Animals weigh choices, absorb information, learn which behavior yields rewards, and solve problems intelligently, and they do all of this within a framework of natural tendencies that have proven their value over the ages. Genes are definitely part of the equation, but to say that animals are nothing but machines controlled by genes is like saying that a Rembrandt is nothing but fabric and paint, or that a brain is a mere collection of neurons. While not incorrect, such statements miss by a mile the higher levels of organization.

Returning to our mother dog, it is easy to recognize in her behavior a complex psychology shaped by a long history of reliance on maternal care. The tendency to feed and clean dependent young is well established for excellent reasons. At the same time, the entrenched nature of the tendency makes it vulnerable to exploitation, as when people gave the dog tiger cubs to raise. Not that this matters much to the mother. From an evolutionary perspective, care for non-offspring may be maladaptive, but from a psychological perspective, it remains entirely authentic and fitting behavior for the species. Another dog, at Beijing Zoo, recently acted as wet nurse for three snow-leopard cubs whose mother had abandoned them.[238]

And so, the dog at the Thai zoo really hadn't done anything unusual, nothing that a good canine wouldn't or shouldn't do. Her behavior did provide a stark reminder, though, of how narrow a portrayal of nature the nearby statue offered. The statue was intended to show selection at work, but could not begin to convey the variety of outcomes evolution has produced. Paradoxically, harsh selection processes have led to some amazingly cooperative species with character traits such as loyalty, trust, sympathy, and generosity.

The Midwife Bat

Before we now conclude that animals and people can be truly unselfish, we need to subject the terms "altruism" and "kindness" to the same scrutiny as was just applied to "selfishness." Here, too, we risk confusion: functional altruism—in which one individual gains from another's actions—does not necessarily rest on intended kindness, in which someone else's well-being is the goal.

When a blue jay gives alarm shrieks for a red-tailed hawk gliding around the corner, does he do so in order to warn others? All potential prey of the hawk take immediate action, and thus profit from the jay's alert, whereas the jay takes enormous risks, telling the hawk, in effect: "Here I am!" On the surface, this seems an act of unmitigated altruism. The critical question remains, however, whether the jay cared about the others: did he even realize the wider impact of his calls?

There exist many examples of altruism in which awareness of what the behavior means to others is questionable. This is especially true for social insects, which sacrifice themselves on a massive scale for their colony and queen. Many other animals help each other find food and water, avoid predation, raise offspring, and so on. Only a few of the largest-brained animals, however, seem to operate with a solid understanding of how their behavior affects others. When these animals go out of their way to help others without any clear benefits for themselves, it is possible that the other's welfare is their goal. I am thinking, for example, of how Binti Jua, the lowland gorilla at Chicago's Brookfield Zoo, scooped up and gently transported an unconscious boy who had fallen into her enclosure. Binti followed a chain of action no one had taught her, resulting in the boy's rescue.[239]

In another incident, a British tourist was protected by dolphins in the Gulf of Akaba off the Red Sea. While cavorting with dolphins, the man was attacked by sharks. When his companions on the vessel heard his screams, they thought at first it was a joke, until they saw blood stain the water. Three dolphins surrounded the injured victim, leaping up and smacking the water with their tails and flippers, and successfully kept the sharks at bay.[240]

In my work on the evolution of morality, I have found many instances of animals caring for one another and responding to others' distress. For example, chimpanzees will approach a victim of attack, put an arm around her and gently pat her back, or groom her. These reassuring encounters, termed *consolations*, are so predictable that my students and I have recorded hundreds of instances.[241] In monkeys, on the other hand, consolation has never been demonstrated. On the contrary, monkeys often avoid victims of aggression. Our closest relatives, the anthropoid apes, thus seem more empathic than monkeys. Apes may be able to perceive the world from someone else's perspective, and hence understand what is wrong with the other, or what the other needs.

Nadie Ladygina-Kohts noticed similar empathic tendencies in her young chimpanzee, Yoni, whom she raised in Moscow at the beginning of the twentieth century. Kohts, who analyzed Yoni's behavior in the minutest detail, discovered that the only way to get him off the roof of her house (much more effectively than by holding out a reward) was to appeal to his feelings of concern for her:

> If I pretend to be crying, close my eyes and weep, Yoni immediately stops his plays or any other activities, quickly runs over to me, all excited and shagged, from the most remote places in the house, such as the roof or the ceiling of his cage, from where I could not drive him down despite my persistent calls and entreaties. He hastily runs around me, as if looking for the offender; looking at my face, he tenderly takes my chin in his palm, lightly touches my face with his finger, as though trying to understand what is happening, and turns around, clenching his toes into firm fists.[242]

In previous books, such as *Good Natured* (1996), I have amassed other examples in support of this empathic capacity in the chimpanzee and its closest relative, the bonobo. For instance, an adult daughter brought fruit down from a tree to her aging mother, who was too old to climb. In another instance, juveniles interrupted their rambunctious play each time they got close to a terminally sick companion. There is also the report of an old male leading a blind female around by the hand, and of an ape who released a damaged bird by climbing to the highest point of a tree, spreading the bird's wings, and sending it off through the air. This individual seemed to have an idea of what kind of assistance might be best for an injured bird. There exist ample stories of this sort about apes that suggest a capacity to assist others insightfully.

But even though apes may be special in this regard, we cannot exclude similar capacities in other animals. A well-documented instance of possible altruism concerns a very different species: Rodrigues fruit bats in a breeding colony in Florida, studied by Thomas Kunz, a biologist at the University of Boston.[243] By chance, Kunz witnessed an exceptionally difficult birthing process in which a mother bat failed to adopt the required feet-down position. Instead, she continued to hang upside-down. Taking on a midwife role, another female spent no less than two and a half hours assisting the inexperienced mother. She licked and groomed her behind, and wrapped her wings around her, perhaps so as to prevent the emerging pup from falling. She also repeatedly fanned the exhausted

*A Rodrigues fruit bat is giving birth, hanging in the correct
feet-down position, perhaps mimicking the helper female,
on the left, who adopted this position several times in front
of her when the pregnant female failed to do so on her own.
(Drawing by Thomas Kunz, with kind permission).*

mother with her wings. But what amazed the biologist most
was that the helper seemed to be *instructing* the mother: the
mother adopted the correct feet-down position only after the
helper had done so right in front of her. On four separate occasions the helper adopted the correct position in full view of

the mother—a position normally used only for urination or defecation, which the helper didn't engage in—and each time the mother followed the helper's example.

It looked very much as if the midwife bat was aware of the difficulties the mother's unorthodox position was causing, and that she tutored the mother to do the right thing. If she indeed monitored the effects of her actions and deliberately strove for a successful delivery, the helper's behavior was not just functionally but also intentionally altruistic. When the pup was finally born, it climbed onto its mother's back assisted by head-nuzzling from the helper female.

We easily recognize such helping tendencies, because they are prominent in our own species. This is abundantly clear when people crawl into smoking ruins to save others, such as during earthquakes and fires. Given our talent for risk assessment, there can be nothing inadvertent about such behavior. When Lenny Skutnik dove into the icy Potomac River in Washington, D.C., to rescue a plane-crash victim, or when European civilians sheltered Jewish families during World War II, incredible risks were taken on behalf of complete strangers. Even if reward comes afterward in the form of a medal or a moment on the evening news, this is of course never the motive. No sane person would willingly risk his life for a piece of metal or five minutes of televised glory. The decision to help is instantaneous and impulsive, without much time to think. When fugitives knock on the door, one determines there and then whether to take them in.

But even if many heroic acts escape traditional biological explanations in terms of "short-sighted selfishness," this doesn't make the underlying tendencies counterevolutionary. More than likely, the helping responses of dolphins, gorillas, or people toward strangers in need evolved in the context of a close-knit group life in which most of the time such actions benefited relatives and companions able to repay the favor. The impulse to help was therefore never totally without survival value to the one showing the impulse. But, as so often, the impulse became dissociated from the consequences that shaped its evolution, which permitted it to be expressed even when payoffs were unlikely. The impulse thus was emancipated to the point where it became genuinely unselfish.

Depressed Rescue Dogs

The animal literature is filled with examples of normal behavior under unusual circumstances. Followed by a single file of goslings, Konrad Lorenz demonstrated the tendency of these birds to imprint on the first moving object they lay their eyes on. He thus permanently confused their sense of species-belonging. Niko Tinbergen saw stickleback fish in a row of tanks in front of his laboratory window, in Leiden, make furious territorial displays at the mail delivery van in the street below. At the time, Dutch mail vans were bright red, the same color as the male stickleback's underbelly during the breeding season, and the fish mistook the van for an intruder of their own species.

Artificial situations sometimes help us see more clearly how behavior is regulated. When goslings do the normal thing, following their mom around all day, one might think that they share our exalted view of motherhood. We are quickly disabused of this notion, however, when they follow a bearded zoologist with equal devotion. And when sticklebacks defend their territory, we might think that they want to keep competitors out, whereas in reality they are only reacting to a species-typical red flag. What animals really are after is not always evident, and tinkering with conditions is a way to find out.

For altruistic behavior, an informative context is that of rescue dogs. Trainers tap into the inborn tendency of these cooperative hunters to come to each other's aid. Time and again, dogs demonstrate this ability spontaneously towards their human "pack members." An example is the occasion on which a rottweiler and a golden retriever crawled side by side on their bellies toward their master, who had broken through the ice on a frozen lake. The heavy man managed to grab their collars, one in each hand, upon which both dogs inched backward, pulling him out.[244]

Rescue dogs are trained to perform such responses on command, often in repulsive situations, such as fires, that they would normally avoid unless the entrapped individuals are familiar. Training is accomplished with the usual carrot-and-stick method. One might think, therefore, that the dogs perform like Skinnerian rats, doing what has been reinforced in the past, partly out of instinct, partly out of a desire for tidbits. If they save human lives, one could argue, they do so for purely selfish reasons.

The image of the rescue dog as a well-behaved robot is hard to maintain, however, in the face of their attitude under trying circumstances with few survivors, such as in the aftermath of the bombing of the Murrah Federal Building in Oklahoma City. When rescue dogs encounter too many dead people, they lose interest in their job regardless of how much praise and goodies they get.

This was discovered by Caroline Hebard, the U.S. pioneer of canine search and rescue, during the Mexico City earthquake of 1985. Hebard recounts how her German shepherd, Aly, reacted to finding corpse after corpse and few survivors. Aly would be all excited and joyful if he detected human life in the rubble, but became depressed by all the death. In Hebard's words, Aly regarded humans as his friends, and he could not stand to be surrounded by so many dead friends: "Aly fervently wanted his stick reward, and equally wanted to please Caroline, but as long as he was uncertain about whether he had found someone alive, he would not even reward himself. Here in this gray area, rules of logic no longer applied."[245]

The logic referred to is that a reward is just a reward: there is no reason for a trained dog to care about the victim's condition. Yet, all dogs on the team became depressed. They required longer and longer resting periods, and their eagerness for the job dropped off dramatically. After a couple of days, Aly clearly had had enough. His big brown eyes were mournful, and he hid behind the bed when Hebard wanted to take him out again. He also refused to eat. All other dogs on the team had lost their appetites as well.

The solution to this motivational problem says a lot about what the dogs wanted. A Mexican veterinarian was invited to act as stand-in survivor. The rescuers hid the volunteer somewhere in a wreckage and let the dogs find him. One after another the dogs were sent in, picked up the man's scent, and happily alerted, thus "saving" his life. Refreshed by this exercise, the dogs were ready to work again.[246]

What this means is that trained dogs rescue people only partly for approval and food rewards. Instead of performing a cheap circus trick, they are emotionally invested. They relish the opportunity to find and save a live person. Doing so also constitutes some sort of reward, but one more in line with what Adam Smith, the Scottish philosopher and father of economics, thought to underlie human sympathy: all that we derive from sympathy, he said, is the pleasure of seeing someone else's fortune. Perhaps this doesn't seem like much, but it means a lot to many people, and apparently also to some big-hearted canines.

Under certain conditions and for certain species, therefore, we can drop the customary quotation marks around "altruism." At least in some cases, we seem to be dealing with the genuine article: a good deed done *and* intended.

Apples and Oranges

It is not hard to see why biologists call the problems they deal with multilayered. At the evolutionary level a behavior may be self-serving; at the psychological level it may be kind and unselfish; and at yet another level it may be best understood by the effect of hormones on certain brain areas. Similarly, from the performer's perspective a behavior may be a mere reflex or fully deliberate, yet this matters little to the recipient, who mainly cares about whether the behavior helps or harms him.

When we freely jump from one level or perspective to another we run the risk of forgetting to keep our language straight. For example, nature documentaries now customarily discuss animal behavior in the shorthand of evolutionary biology ("The croaking frog advertises his genetic superiority to potential mates"), making us forget that animals know nothing about the genetic story. Even worse is that scientists who operate on one level sometimes can't stand another level's idiom, and vice versa. This explains why some flinch at a behavior being called altruistic, whereas others flinch at the same behavior being called selfish. In fact, both may be right within their respective frameworks.

If one biologist's apples are another's oranges, this obviously creates a communication problem. We usually resolve the difficulty by asking whether someone is talking at the "proximate" (direct causation) or "ultimate" (adaptive value) level, but this distinction has never caught on outside of biology.

The tension between the two is forever there, however. The mother dog who raises tiger cubs is at once extraordinarily generous and doing what her genes, based on millions of years of self-service, nudge her to do. By following her natural impulses, she illustrates the contradictions that lend so much richness to evolutionary accounts that we will never be done mining their meaning.

Down with Dualism!

Two Millennia of Debate About Human Goodness

"We approve and we disapprove because we cannot do otherwise. Can we help feeling pain when the fire burns us? Can we help sympathizing with our friends? Are these phenomena less necessary or less powerful in their consequences, because they fall within the subjective sphere of experience?"

Edward Westermarck, 1912

*E*dward Westermarck's writings, including those about his journeys to Morocco, kept me busy as I leaned back in a cushy seat on a jet from Tokyo to Helsinki. More comfortable than a camel, I bet! I was on my way to an international conference in honor of the Swedish-Finn, who lived from 1862 until 1939, and who was the first to bring Darwinism to the social sciences.

His books are a curious blend of dry theorizing, detailed anthropology, and secondhand animal stories. He gives the example of a vengeful camel that had been excessively beaten on multiple occasions by a fourteen-year-old "lad" for loitering or turning the wrong way. The camel passively took the punishment, but a few days later, finding itself unladen and alone on the road with the same conductor, "seized the unlucky boy's head in its monstrous mouth, and lifting him up in the air flung him down again on the earth with the upper part of the skull completely torn off, and his brains scattered on the ground."[247]

I don't know much about camels, but stories of delayed revenge abound in the zoo world, especially about apes and elephants. We now have systematic data on how chimpanzees punish negative actions with other negative actions—a pattern called a "revenge system"—and how if a macaque is attacked by a dominant member of its troop it will turn around to redirect aggression against a vulnerable, younger relative of its attacker.[248] Such behavior falls under what Westermarck called the "retributive emotions," but for him "retributive" went beyond its usual connotation of getting even. It included positive tendencies, such as gratitude and the repayment of services. Depicting the retributive emotions as the cornerstone of human morality, Westermarck weighed in on the question of its origin while antedating modern discussions of evolutionary ethics, which often take the related concept of reciprocal altruism as their starting point.[249]

That Westermarck goes unmentioned in the latest books on evolutionary ethics, or serves only as a historic footnote, is not because he paid attention to the wrong phenomena or held untenable views about ethics, but because his writing conveyed a belief in human goodness. He felt that morality comes naturally to people. Contemporary biologists have managed to banish this view to the scientific fringes under the influence of the two Terrible Toms—Thomas Hobbes and Thomas Henry Huxley—who both preached that the original state of humankind, and of nature in general, is one in which selfish goals are pursued without regard for others. Compromise, symbiosis, and mutualism were not terms the Toms considered particularly useful, even though these outcomes are not hard to come by in both nature and human society.

Are we naturally good? And if not, whence does human goodness come? Is it one of our many marvelous inventions, like the wheel and toilet training, or could it be a mere illusion? Perhaps we are naturally bad, and just pretend to be good?

Every possible answer to these questions has been seriously advocated by one school of thought or another. I myself have struggled with the question of human nature, contrasting the views of present-day biologists—from whom an admission of human virtue is about as hard to extract as a rotten tooth—with the belief of many philosophers and scientists, including Charles Darwin, that our species moderates its selfishness with a healthy dose of fellow-feeling and kindness. Anyone who explores this debate will notice how old it is— including, as it does, explicit Chinese sources, such as Mencius, from before the Western calendar—so that we can justifiably speak of a perennial controversy.

Westermarck Beats Freud

In a stately building on a wintry, dark Helsinki day, not far from his childhood home, we discussed Westermarck's brave Darwinism, which was initially applauded but soon opposed by contemporary big shots such as Sigmund Freud and Claude Lévi-Strauss. Their resistance was so effective that the Finn has been largely forgotten.

His most controversial position concerned incest. Both Freud and many anthropologists were convinced that there would be rampant sex within the human family if it were not for the incest taboo. Freud believed that the earliest sexual excitations and fantasies of children are invariably directed at close family members, while Lévi-Strauss declared the incest taboo the ultimate cultural blow against nature—it was what permitted humanity to make the passage from nature to culture.

These were high-flown notions, which carried the stunning implication that our species was somehow predestined to free itself of its biological shackles. Westermarck didn't share the belief that our ancestors started out with rampant, promiscuous sex over which they gained control only with great difficulty. He instead saw the nuclear family as humanity's age-old reproductive unit, and proposed that early association within this unit (such as normally found between parent and off-spring and among siblings) kills sexual desire. Hence, the desire isn't there to begin with. On the contrary, individuals who grow up together from an early age develop an actual sexual *aversion* for each other. Westermarck proposed this as an evolved mechanism with an obvious adaptive value: it prevents the deleterious effects of inbreeding.

In the largest-scale study on this issue to date, Arthur Wolf, an anthropologist at Stanford University, spent a lifetime examining the marital histories of 14,402 Taiwanese women in a "natural experiment" dependent on a peculiar Chinese marriage custom. Families used to adopt and raise little girls as future daughters-in-law. This meant that they grew up since early childhood with the family's son, their intended husband. Wolf compared the resulting marriages with those arranged between men and women who did not meet until their wedding day. Fortunately for science, official household registers were kept during the Japanese occupation of Taiwan. These registers provide detailed information on divorce rates and number of children, which Wolf took as measures of, respectively, marital happiness and sexual activity. His data supported the Westermarck effect: association in the first years of life appears to compromise marital compatibility.[250]

These findings are especially damaging to Freud, because if Westermarck is right then Oedipal theory is wrong. Freud's

thinking was premised on a supposed sexual attraction between members of the same family that needs to be suppressed and sublimated. His theory would predict that unrelated boys and girls who have grown up together will marry in
absolute bliss, as there is no taboo standing in the way of their
primal sexual desires. In reality, however, the signs are that
such marriages often end in misery. Co-reared boys and girls
resist being wed, arguing that they are too much like brother
and sister. The father of the bride sometimes needs to stand
with a stick by the door during the wedding night to prevent
the two from escaping the situation. In these marriages, sexual
indifference seems to be the rule, and adultery a common out-
let. As Wolf exclaimed at the conference, Westermarck may
have been less flamboyant, less self-assured, and less famous
than any of his mighty opponents; the fundamental difference
was that he was the only one who was right!

A second victim is Lévi-Stauss, who built his position entirely on the assumption that animals lead disorderly lives in
which they do whatever they please, including committing incest. We now believe, however, that monkeys and apes are
subject to exactly the same inhibitory mechanism as proposed
by Westermarck. Many primates prevent inbreeding through
migration of one sex or the other. The migratory sex meets
new, unrelated mates, while the resident sex gains genetic diversity by breeding with immigrants. In addition, close kin
who stay together avoid sexual intercourse. This was first observed in the 1950s by Kisaburo Tokuda in a group of Japanese

macaques at the Kyoto Zoo. A young adult male who had risen to the top rank made full use of his sexual privileges, mating frequently with all of the females except for one: his mother.[251] This was not an isolated case; mother-son matings are strongly suppressed in all primates. Even in the sexy bonobos, this is the one partner combination in which intercourse is rare or absent. Observation of thousands of matings in a host of primates, both captive and wild, has demonstrated the suppression of incest.

The Westermarck effect serves as a showcase for Darwinian approaches to human behavior because it so clearly rests on a *combination* of nature and nurture: it has a developmental side (learned sexual aversion), an innate side (the way early familiarity affects sexual preference), a cultural side (some cultures raise unrelated children together and others raise siblings of the opposite sex apart, but most have family arrangements that automatically lead to sexual aversion among relatives), a likely evolutionary reason (suppression of inbreeding), and direct parallels with animal behavior. On top of this comes the cultural *taboo*, which is unique for our species. An unresolved issue is whether the taboo merely serves to formalize and reinforce the Westermarck effect or adds a substantially new dimension.

That Westermarck's integrated view was underappreciated at the time is understandable, as it flew in the face of the Western dualistic tradition. What is less understandable is why these dualisms remain popular today. Westermarck was more

Darwinian than some contemporary evolutionary biologists, who are best described as Huxleyan.

Bulldog Bites Master

In 1893, before a large audience in Oxford, Huxley publicly tried to reconcile his dim view of the nasty natural world with the kindness occasionally encountered in human society. Huxley realized that the laws of the physical world are unalterable. He felt, however, that their impact on human existence could be softened and modified if people kept nature under control. Comparing us with the gardener who has a hard time keeping weeds out of his garden, he proposed ethics as humanity's cultural victory over the evolutionary process.[252]

This was an astounding position for two reasons. First, it deliberately curbed the explanatory power of evolution. Since many people consider morality the essence of our species, Huxley was in effect saying that what makes us human is too big for the evolutionary framework. This was a puzzling retreat by someone who had gained a reputation as "Darwin's Bulldog" owing to his fierce advocacy of evolutionary theory. The solution that Huxley proposed was quintessentially Hobbesian in that it stated that people are fit for society only by education, not nature.

Second, Huxley offered no hint whatsoever where humanity could possibly have unearthed the will and strength to go against its own nature. If we are indeed born competitors who don't care one bit about the feelings of others, how in the world did we decide to transform ourselves into model citizens? Can people for generations maintain behavior that is out of character, like a bunch of piranhas who decide to become vegetarians? How deep does such a change go? Are we the proverbial wolves in sheep's clothing: nice on the outside, nasty on the inside? What a contorted scheme!

It was the only time Huxley visibly broke with Darwin. As aptly summarized by Huxley's biographer, Adrian Desmond: "[He] was forcing his ethical Ark against the Darwinian current which had brought him so far."[253] Two decades earlier, in *The Descent of Man*, Darwin had stated the continuity between human nature and morality in no uncertain terms. The reason for Huxley's departure has been sought in his suffering at the cruel hand of nature, which had just taken his beloved daughter's life, and in his need to make the ruthlessness of the Darwinian cosmos palatable to the general public. He could do so, he felt, only by dislodging human ethics, declaring it a cultural innovation.

This dualistic outlook was to get an enormous respectability boost from Freud's writings, which throve on contrasts between the conscious and subconscious, the ego and superego, Eros and Death, and so on. As with Huxley's gardener and garden, Freud was not just dividing the world in symmetrical halves: he saw struggle everywhere! He explained the incest taboo and other moral restrictions as the result of a violent break with the freewheeling sexual life of the primal horde, culminating in the collective slaughter of an overbearing father by his sons. And he let civilization arise out of a renunciation of instinct, the gaining of control over the forces of nature, and the building of a cultural superego. Not only did he keep animals at a distance, his view also excluded women. It was the men who reached the highest peaks of civilization, carrying out tortuous sublimations "of which women are little capable."[254]

Humanity's heroic combat against forces that try to drag us down remains a dominant theme within biology today. Because of its continuity with the doctrine of original sin, I have characterized this viewpoint as "Calvinist sociobiology."[255] Let me offer a few illustrative quotations from today's two most outspoken Huxleyans.

Declaring ethics a radical break with biology, and feeling that Huxley had not gone far enough, George Williams has written extensively about the wretchedness of Mother Nature. His stance culminates in the claim that human morality is an inexplicable accident of the evolutionary process: "I account for morality as an *accidental* capability produced, in its boundless *stupidity*, by a biological process that is normally opposed to the expression of such a capability" (my italics). In a similar vein, Richard Dawkins has declared us "nicer than is good for our selfish genes," and warns that "we are never allowed to forget the narrow tightrope on which we balance above the Darwinian abyss." In a recent interview, Dawkins explicitly endorsed Huxley: "What I am saying, along with many other people, among them T. H. Huxley, is that in our political and social life we are entitled to throw out Darwinism, to say we don't want to live in a Darwinian world."[256]

Poor Darwin must be turning in his grave, because the world implied here is totally unlike what he himself envisioned. Again, what is lacking is an indication of how we can possibly negate our genes, which the same authors at other times don't hesitate to depict as all-powerful. Thus, first we are told that our genes know what is best for us, that they control our lives, programming every little wheel in the human survival machine. But then the same authors let us know that we have the option to rebel, that we are free to act differently. The obvious implication is that the first position should be taken with a grain of salt.

Like Huxley, these authors want to have it both ways: human behavior is an evolutionary product except when it is hard to explain. And like Hobbes and Freud, they think in dichotomies: we are part nature, part culture, rather than a well integrated whole. Their position has been echoed by popularizers such as Robert Wright and Matt Ridley, who say that virtue is absent from people's hearts and souls, and that our species is potentially but not naturally moral.[257] But what about the many people who occasionally experience in themselves and others a degree of sympathy, goodness, and generosity? Wright's answer is that the "moral animal" is a fraud: "[T]he pretense of selflessness is about as much part of human nature as is its frequent absence. We dress ourselves up in tony moral

language, denying base motives and stressing our at least minimal consideration for the greater good; and we fiercely and
self-righteously decry selfishness in others."[258]

To explain how we manage to live with ourselves despite
this travesty, theorists have called upon self-deception and denial. If people think they are at times unselfish, so the argument goes, they must be hiding the selfish motives from themselves. In other words, all of us have two agendas: one hidden
in the recesses of our minds, and one that we sell to ourselves
and others. Or, as philosopher Michael Ghiselin concludes,
"Scratch an 'altruist,' and watch a 'hypocrite' bleed." In the ultimate twist of irony, anyone who doesn't believe that we are
fooling ourselves, who feels that we may be genuinely kind, is
called a wishful thinker and thus stands accused of fooling
himself![259]

This entire double-agenda idea is another obvious
Freudian scheme. And like a UFO sighting, it is unverifiable:
hidden motives are indistinguishable from absent ones. The
quasi-scientific concept of the subconscious conveniently
leaves the fundamental selfishness of the human species intact despite daily experiences to the contrary.[260] I blame
much of this intellectual twisting and turning on the unfortunate legacy of Huxley, about whom evolutionary biologist
Ernst Mayr didn't mince any words: "Huxley, who believed in
final causes, rejected natural selection and did not represent
genuine Darwinian thought in any way. . . . It is unfortunate,
considering how confused Huxley was, that his essay [on evo-

lutionary ethics] is often referred to even today as if it were authoritative."[261]

Moral Emotions

Westermarck is part of a long lineage, going back to Aristotle and Thomas Aquinas, which firmly anchors morality in the natural inclinations and desires of our species. Compared to Huxley's, his is a view uncompounded by any need for invisible agendas and discrepancies between how we are and how we wish to be: morality has been there from the start. It is part and parcel of human nature.

Emotions occupy a central role in that, as Aristotle said, "Thought by itself moves nothing." Modern cognitive psychologists and neuroscientists confirm that emotions, rather than being the antithesis of rationality, greatly aid thinking. They speak of emotional intelligence. People can reason and deliberate as much as they want, but if there are no emotions attached to the various options in front of them, they will never reach a decision or conviction.[262] This is critical for moral choice, because if anything, morality involves strong convictions. These don't—or rather can't—come about through a cool Kantian rationality; they require caring about others and powerful gut feelings about right and wrong.

Westermarck discusses, one by one, a whole range of what philosophers before him used to call the "moral sentiments." He classifies the retributive emotions into those derived from resentment and anger, which seek revenge and punishment, and those that are more positive and prosocial. Whereas in his time there were few good animal examples of the moral emotions—hence his occasional reliance on Moroccan camel stories—we know now that there are many parallels in primate behavior. Thus, he discusses "forgiveness," and how the turning of the other cheek is a universally appreciated gesture: we now know from our studies that chimpanzees kiss and embrace and that monkeys groom each other after fights.[263] Westermarck sees protection of others against offenders resulting from "sympathetic resentment"; again, this is a common pattern in monkeys and apes, and in many other animals, who stick up for their friends, defending them against attackers. Similarly, the retributive kindly emotions ("desire to give pleasure in return for pleasure") have an obvious parallel in what biologists now label reciprocal altruism, such as providing assistance to those who assist in return.[264]

When I watch primates, measuring how they share food in return for grooming, comfort victims of aggression, or wait for the right opportunity to get even with a rival, I see very much the same emotional impulses that Westermarck analyzed. A group of chimpanzees, for example, may whip up an outraged chorus of barks when the dominant male overdoes his punishment of an underling, and in the wild they form cooperative hunting parties that share the spoils of their efforts. Although I shy away from calling chimpanzees "moral beings," their psychology contains many of the ingredients that, if also present in the progenitor of humans and apes, must have allowed our ancestors to develop a moral sense. Instead of seeing morality as a radically new invention, I tend to view it as a natural outgrowth of ancient social tendencies.

Westermarck was far from naïve about how morality is maintained; he knew it required both approval and negative sanctions. For example, reflecting on an issue that today we might relate to developments taking place in South Africa's Truth and Reconciliation Commission, he explains how forgiveness prohibits revenge but not punishment. Punishment is a necessary component of justice, whereas revenge—if let loose—only destroys. Like Adam Smith before him, Westermarck recognized the moderating role of sympathy: "The more the moral consciousness is influenced by sympathy, the more severely it condemns any retributive infliction of pain which it regards as undeserved."[265]

The most insightful part of his writing is perhaps where Westermarck tries to come to grips with what defines a moral emotion as moral. Here he shows that there is much more to these emotions than raw gut feeling. In analyzing these feelings he introduces the notion of "disinterestedness." Emotions, such as gratitude and resentment, directly concern one's own interests—how one has been treated or how one wishes to be treated—and hence are too egocentric to be moral. Moral emotions, in contrast, are disconnected from one's immediate situation: they deal with good and bad at a more abstract, disinterested level. It is only when we make general judgments of how *anyone* ought to be treated that we can begin to speak of moral approval and disapproval. This is an area in which humans go radically farther than other primates.[266]

Westermarck was ahead of his time, and he went well beyond Darwin's thinking on these matters. In spirit, however, the two were on the same line. Darwin believed that there was plenty of room within his theory to accommodate the origins of morality, and he attached great importance to the capacity for sympathy. He by no means excluded animals from this view: "Many animals certainly sympathize with each other's distress or danger."[267] He has been proven right; laboratory experiments on monkeys and even rats have shown powerful vicarious distress responses. The sight of a conspecific in pain or trouble often calls forth a reaction to ameliorate the situation. These reactions undoubtedly derive from parental care, in which vulnerable individuals are tended with great care, but in many animals they stretch well beyond this situation, including relations among unrelated adults.[268]

Darwin did not see any conflict between the harshness of the evolutionary process and the gentleness of some of its products. As discussed in the previous chapter with regard to the distinction between motive and function, all one needs to do is make a distinction between how evolution operates and the actual psychologies it has produced. Darwin knew this better than anyone, expressing his views most clearly when he emphasized continuity with animals even in the moral domain. In *The Descent of Man*, he takes exactly the opposite position of those who, like Huxley, view morality as a violation of evolutionary principles: "Any animal whatever, endowed with well-marked social instincts, the parental and filial affections being here included, would inevitably acquire a moral sense or conscience, as soon as its intellectual powers had become as well developed, or nearly as well developed, as in man."[269]

The *Ke* Willow

There is never much new under the sun. Westermarck's emphasis on the retributive emotions, whether friendly or vengeful, reminds one of Confucius' reply to the question whether there is any single word that may serve as prescription for all of one's life. Confucius proposed "reciprocity" as such a word. Reciprocity is also, of course, the crux of the Golden Rule ("Do unto others as you would have them do unto you"), which remains unsurpassed as a summary of human morality.

A follower of the Chinese sage, Mencius, wrote extensively about human goodness during his life, from 372 to 289 B.C.[270] Mencius lost his father when he was only three, and his mother made sure he received the best possible education. The mother is at least as well known as her son, and still serves as a maternal model to the Chinese for her absolute devotion.

Called the "second sage" because of his great influence, Mencius had a revolutionary bent in that he stressed the obligation of rulers to provide for the common people. Recorded on bamboo clappers and handed down to his descendants and their students, his writings show that the debate about whether we are naturally moral, or not, is ancient indeed. In one exchange, Mencius reacts against Kaou Tsze's views, which are strikingly similar to Huxley's gardener and garden metaphor: "Man's nature is like the *ke* willow, and righteousness is like a cup or a bowl. The fashioning of benevolence and righteousness out of man's nature is like the making of cups and bowls from the *ke* willow."[271]

Mencius replied:

> Can you, leaving untouched the nature of the willow, make with it cups and bowls? You must do violence and injury to the willow, before you can make cups and bowls with it. If you must do violence and injury to the willow, before you can make cups and bowls with it, *on your principles* you must in the same way do violence and injury to humanity in order to fashion from it benevolence and righteousness! Your words alas! would certainly lead all men on to reckon benevolence and righteousness to be calamities.

Evidently, the origins of human kindness and ethics were a point of debate in the China of two millenia ago. Mencius believed that humans tend toward the good as naturally as water flows downhill. This is also evident from the following remark, in which he seeks to exclude the possibility of a double agenda on the grounds that the moral emotions, such as sympathy, leave little room for this:

> When I say that all men have a mind which cannot bear to see the suffering of others, my meaning may be illustrated thus: even nowadays, if men suddenly see a child about to fall into a well, they will without exception experience a feeling of alarm and distress. They will feel so, not as a ground on which they may gain the favor of the child's parents, nor as a ground on which they may seek the praise of their neighbors and friends, nor from a dislike to the reputation of having been unmoved by such a thing. From this case we may perceive that the feeling of commiseration is essential to man.

Mencius' example is strikingly similar to both the one by Westermarck ("Can we help sympathize with our friends?") and Smith's famous definition of sympathy ("How selfish soever man may be supposed to be . . ."). The central idea underlying all three statements is that distress at the sight of another's pain is an impulse over which we exert no control: it grabs us instantaneously, like a reflex, leaving us without the time to weigh the pros and cons. Remarkably, all of the alternative motives that Mencius considers occur in the modern literature, usually under the heading of reputation building. The big difference is, of course, that Mencius rejects these explanations as too contrived given the immediacy and force of the sympathetic response. Manipulation of public opinion is entirely possible at other times, he says, but not at the moment a child falls into a well.

I couldn't agree more. Evolution has produced species that follow genuinely cooperative impulses. I don't know whether people are, deep down, good or evil, but I do know that to believe that each and every move is selfishly calculated overestimates human mental powers, let alone those of other animals.[272]

Interesting additional evidence comes from child research. Freud, B. F. Skinner, and Jean Piaget all believed that the child learns its first moral distinctions through fear of punishment and a desire for praise. Like Huxleyan biologists who see morality as culturally imposed upon a nasty human nature, they conceived morality as coming from the outside, imposed by adults upon a passive, naturally selfish child. Children were thought to adopt parental values to construct a superego, the moral agency of the self. Left to their own devices, like the children in William Golding's *Lord of the Flies*, they would never arrive at anything even close to morality.

Already at an early age, however, children know the difference between moral principles ("Do not steal") and cultural conventions ("No pajamas at school"). They apparently appreciate that the breaking of certain rules distresses and harms others, whereas the breaking of other rules merely violates expectations about what is appropriate. Their attitudes don't seem to be based purely on reward and punishment. Whereas pediatric handbooks still depict young children as self-centered monsters, we know now that by one year of age they spontaneously comfort people in distress, and that soon thereafter they begin to develop a moral perspective through interactions with other members of their species.[273]

Rather than being nicer than is good for our genes, we may be just nice enough. Thus, the child is not going against its own nature by developing a caring, moral attitude, and civil society is not like an out-of-control garden subdued by a sweating gardener. We are merely following evolved tendencies

How refreshingly simple!

Epilogue

The Squirrel's Jump

Will we ever manage to fit both the ape and the sushi master in the same family picture? The ape represents our natural, primitive side, which we sometimes use as a caricature of ourselves just to show how far we have come. The sushi master epitomizes human sophistication, artistry, and know-how. We eat the fugu (blowfish sushi) trusting the chef's skills, which he learned from other chefs, and they in turn from those before them.[274] How can these two different versions of ourselves—unvarnished nature and cultural refinement—ever be reconciled?

There is a long lineage of thinkers that has never had a problem with it. They carry their talismans, derived from a sense of kinship with animals, proudly around their necks, and rarely bother to emphasize what separates us. Not that they are oblivious to the differences, but their first goal is to understand humanity in the wider context of nature. This is the Darwistotelian view, according to which humans have both of their feet firmly planted on this earth, which brought them forth in every imaginable sense. No area of human behavior is exempted: we arrived as a single package produced by exactly the same forces that produced all slightly different looking packages around us.

Then there is the equally ancient school that assigns us a special spot in the universe while sternly warning against attempts, whether expressed in anthropomorphism or general assumptions, to blur the line between ourselves and other creatures. In this view, the sushi master and the ape are fully decoupled, if not in body then at least in mind. Actually, humans are the only ones considered as having a mind. Whereas the surrounding world is brutish, mechanical, and amoral, we are blessed with a free will and the ability to direct our societies any way we want. The fact that we share characteristics with animals is no problem, because we have the ability to kick our own nature into shape until it fits civilized society.

When divine sparks fell out of fashion, the widely accepted key to our special success became culture. It was culture that let us push the envelope, break out of it, and start a new life totally unlike the ape's. Culture became a magic, reified concept disconnected from and even antithetical to nature. Culture was seen as something that we produce at will, yet that at the same time produces us. No matter the monumental circularity of this argument, it soon permeated all of the social sciences and humanities. It even won over an occasional biologist. Culture became the escape clause whenever the contract with nature seemed too constraining.

No wonder there is so much animosity behind the culture wars ignited by recent animal behavior discoveries. If culture can't be claimed as uniquely human, where does this leave the second school of thought? It is plain which side I am on, and even if I have covered a great variety of topics in this book—from painting apes to puritanism, and from the role of theory in science to potato-washing monkeys—I have kept my eyes on the question of how knowledge about animal culture influences our self-perception and how whether we grant animals culture is ultimately a human cultural question. It can hardly be coincidental that the push for cultural studies on animals initially came from outside of the intellectual war zone just described—that is, from primatologists untrained in the sharp dualisms of the West.

Now, let me make clear that the issue at hand is not merely a matter of whether the glass is half-full or half-empty. It is not that some scholars emphasize continuity whereas others emphasize difference, and that all we need is to appreciate each other's positions. The implications of one position or the other are far too fundamental. It boils down to the choice between whether we are naturally or artificially moral, or whether or not we are the only "self-made" species on earth. If a growing number of scientists argue that animals rely for their survival on socially transmitted habits and knowledge, and that their strategies vary from group to group, the whole idea of a recent transition from nature to culture is put into question. Not only that: the term "transition" requires reconsideration.

Human uniqueness claims are a bit like advertisements for squirrel-proof bird feeders. I have yet to find a single feeder that stands up to the American gray squirrels in my backyard without being so convoluted that it scares off the birds. Similarly, any claim about human uniqueness is eventually assaulted by an army of scientists who gnaw little holes in it, climb a pole everyone considered too slippery, or make the one impossible jump. The claim that animals have culture may come across to some scholars as precisely such a jump, but it has been made. We can yell at the squirrel as much as we want; we know it will be back.

Where does all this leave us? I see little life left in the position that we humans fall outside of nature, and that it is culture that sets us apart. A retreat to "symbolic culture" as the hallmark of humanity may provide some relief, but in the long run I see a much more fruitful challenge for scholars in search of typically human accomplishments. The time has come to define the human species against the backdrop of the vast common ground we share with other life forms. Instead of being tied to how we are unlike any animal, human identity should be built around how we are animals that have taken certain capacities a significant step farther. We and other animals are both similar and different, and the former is the only sensible framework within which to flesh out the latter.

And so the ape and the sushi master do fit in the same picture, both having learned from others how to process food, and what to eat or not to eat. Even though the ape has none of the symbols surrounding the chef's job, he has come to depend to such a degree on handed-down knowledge that we can safely call both of them cultured. And not only them: the world is chock-full of feathered and furry animals that learn their life's lessons, habits, and songs from one another. With so many cultural creatures surrounding us it is indeed time to carry a few familiar dichotomies to their grave.

NOTES

Prologue

1. Austin (1974).
2. Morris and Morris (1966, p. 102).
3. Kummer (1971).
4. Julien-Joseph Virey (1817), from before the divide between nature and culture, put it as follows: "Nothing falls outside of nature, nothing can escape it. Civil and moral laws, history, the actions of men are merely actions of an animal species subject to the laws of nature" [my translation from French]. In contrast, most of the twentieth century has seen a concept of culture as something with a life of its own: culture creates more culture without any connection with the biological substrate that forms our bodies and minds. Wilson (1998) compared this view of humanity—human on the outside, alien on the inside—to the protagonists in the movie *Invasion of the Body Snatchers.*
5. In line with a theory explained in "The Last Rubicon" (Chapter 6), according to which social learning rests on identification with and the desire to be like a role model, it has been shown that young kittens copy the actions of their mothers more readily than those of unfamiliar female cats (Chelser, 1969).
6. Cheney and Seyfarth (1990).
7. Mineka et al. (1984).
8. Curio (1978) conducted experiments in which one bird mobbed a stuffed owl at the same time that a second bird in a nearby cage was shown another, nonpredatory bird model. Hearing the first bird's mobbing calls, the second bird would react by mobbing the nonpredator. After one such experience, this bird would treat the nonpredator as an enemy whenever it saw

it, and pass its alarm on to other birds. With this ingenious method, "cultural prejudices" could also be created for inanimate objects, such as a bottle of laundry detergent. It shows that the predator image is not inborn but socially learned, making for a flexible way of transmitting knowledge about new dangers in the environment.

9. Galef (1982).

10. Kellogg and Kellogg (1933, p. 141).

11. Custance et al. (1995).

12. Myowa-Yamakoshi and Matsuzawa (1999).

13. Tomasello (1999).

14. The accrual of behavioral modifications toward ever greater complexity has been observed, for example, in the potato washing and wheat sluicing of Japanese macaques on Koshima Island, suggesting that if one follows cultural learning long enough ratcheting can be found in animals as well (Watanabe 1994).

15. Savage-Rumbaugh and Lewin (1994).

1. The Whole Animal: Childhood Talismans and Excessive Fear of Anthropomorphism

16. Cenami Spada (1997).

17. De Waal and Berger (2000) demonstrated that brown capuchin monkeys share more food with a partner who has helped them secure the food by pulling a heavy tray than with partners whose help was unneeded. This experiment was part of a series on social reciprocity, or tit for tat, and mental record keeping of given and received favors in chimpanzees and capuchins. Ultimately, it relates to the cooperative hunting observed in both of these primates in the field. In the hunt, several partners work together but only one captures the prey. Willingness of this individual to share with its helpers may be a prerequisite for continued cooperation.

18. The discovery of the pecking order and other historical details related here come from an interview by John Price with Dag Schjelderup-Ebbe, the sixty-year-old son of Thorleif, published in 1995 in the *Human Ethology Bulletin* 10 (1): 1–6.

19. Freud (1913).

20. Shepard (1996, p. 88).

21. A recent example is Budiansky (1998), who denies animals even the most basic forms of cognition, arguing—in so many words—that people are

so smart they fail to grasp how dumb animals are. He bases this dualism on the absence of language in animals ("We have a terrific piece of software that they simply do not"). Given that neuroscientists believe core consciousness to be entirely nonverbal (Damasio, 1999), it is doubtful, however, that language is the key issue when comparing human and animal minds.

22. The prize for the most amusing cultural reflection of this kind goes to Bertrand Russell (1927), who commented that animals display the national characteristics of the observer: "Animals studied by Americans rush about frantically, with an incredible display of hustle and pep, and at last achieve the desired result by chance. Animals observed by Germans sit still and think, and at last evolve the solution out of their inner consciousness. "

23. Lorenz (1981).

24. Moore and Stuttard (1979).

25. Quoted in Bailey (1986).

26. Garcia et al. (1966).

27. The idea that brains become merely bigger and faster in the course of evolution is still surprisingly widespread despite evidence for substantial structural variation. We know now, for example, that humans and apes possess particular spindle-shaped brain cells in the anterior cingulate cortex that are not found in any other animals. These neurons have been implied in higher cognitive functions typical of humans and apes (Nimchinsky et al. , 1999).

28. Hollard and Delius (1982) and Balda and Kamil (1989).

29. Gallup (1970).

30. The original pigeon "self-recognition" study was conducted by Epstein et al. (1981). Ironically, the failure of attempts at replication led to the reproach that the wrong *strain* of pigeon had been used. Coming from the same scientists who consider entire animal families interchangeable, this was a peculiar complaint. In any case, also when the very same strain of pigeon and an exact copy of the test chamber were used, the team of replicators never got their birds to peck at themselves in front of a mirror (Thompson and Contie, 1994).

31. Heyes (1995) and Povinelli (1997).

32. Serpell (1996).

33. The Athens-Pittsburgh Symposium in the History & Philosophy of Science & Technology, entitled The Problem of Anthropomorphism in Science and Philosophy, was held in May, 1996, in Delphi, Greece.

34. Morgan (1894).

35. Lloyd Morgan's rider went as follows: "To this, however, it should be added, lest the range of the principle be misunderstood, that the canon by no means excludes the interpretation of a particular activity in terms of the higher processes if we already have independent evidence of the occurrences of these higher processes in the animal under observation" (Morgan 1903). For the view that Morgan in fact had nothing against anthropomorphism, see Thomas (1998) and Sober (1998).

36. Kennedy (1992). For an antidote see the volume by Mitchell et al. (1997).

37. This position draws upon the familiar homology argument. Cross-specific similarities in behavior are either "analogies" (independently derived) or "homologies" (owing to shared descent), and the latter is more likely the more closely related the species are. De Waal (1991) discusses evolutionary (as opposed to cognitive) parsimony.

38. Hume (1739, p. 226).

39. Vicchio (1986).

40. Quotes are from Roberts (1996). The author's idea that the horse's chewing movements refer to grazing is not far removed from the ethological concept of ritualization. Evolution has turned many an instrumental act (such as preening or feeding) into a communication signal through exaggeration and increased stereotypy.

41. Nagel (1974).

42. Vermeij (1996), a blind biologist, writes: "If I had difficulty adjusting to blindness, the memory has faded. Almost immediately . . . I discovered the value of echoes for telling me where I was. Sounds bouncing off obstructions provided cues to the size of the room, the position of a tree, the speed of a car, the presence of a person, whether a door was open or closed, and much more. " Atkins (1996) exposes the limitations of Nagel's (1974) question.

43. Similarly, Batson et al. (1990) investigated human response patterns associated with two kinds of empathy: one based on imagining how you would feel in the other person's situation, the other based on imagining how the other feels.

44. Burghardt (1985).

45. A videotape of the incident (and a series of stills in *Stern*, September 5, 1996) shows Binti sitting down upright on a log in a stream while correctly positioning the unconscious boy, cradling him in her lap. It seems as if she is trying to put him on his feet. The Brookfield gorillas might not have

reacted the same to an adult person (i. e. , they probably recognized the boy as a youngster), and they certainly would not have reacted this way to a sack of flour. They would probably have been afraid of the sack at first, but then have opened it, causing a mess (Jay Peterson, curator at the Brookfield Zoo, personal communication).

46. Systematic data on the consolation of distressed individuals by chimpanzees has been provided by de Waal and Aureli (1996). For other accounts of empathy by apes, see de Waal (1996a). For example, in the Arnhem chimpanzee colony a mother put the normal preference for her younger offspring aside when her older offspring was seriously hurt in a scuffle. Ignoring the noisy protests of her infant, she took tender care of this juvenile for weeks until his injuries had healed.

47. Arnhart (1998) explains that Aristotle knew about apes—he had dissected primates and believed that they represented an intermediate form between man and the quadrupeds. One prominent biologist, J. A. Moore (1993), has declared all of biology a footnote to Aristotle.

48. The distinction goes back to the old one between *Naturwissenschaften* (natural sciences) and *Geisteswissenschaften* (sciences of the mind and human spirit), with psychology increasingly adopting the methods and rigor of the first but tracing its intellectual history to the second.

49. Van Iersel was another Dutch ethologist. What Baerends calls "irrelevant behavior" is a humorous reference to so-called displacement activities (such as scratching one's head), which ethologists interpret as a sign of contradictory motivations. The quotation is from Baerends' unpublished lecture at the 1989 International Ethological Conference in Utrecht, the Netherlands.

50. Hodos and Campbell (1969) and Beach (1950).

51. Greenberg and Haraway (1998).

52. Hence Wilson's (1998) offhand comment: "Sociobiology (or Darwinian anthropology, or evolutionary psychology, or whatever more politically acceptable term one chooses to call it) offers a key link in the attempt to explain the biological foundation of human nature. " The reason I stubbornly keep calling myself an ethologist/zoologist despite the many name changes is that I consider the groundbreaking theoretical developments of the 1960s and 1970s a logical continuation of the original ethological agenda, spelled out in Tinbergen's (1963) four research aims: causation, ontogeny, adaptive value, and evolution.

2. The Fate of Gurus
When Silverbacks Become Stumbling Blocks

53. My translation from Zimen, a Lorenz follower and wolf expert well known in Germany (Erik Zimen erzählt . . . , *Wildlife Observer* 12/1999: 93–95).

54. Matricide may be less common in science, but anthropologists dealing with the legacy of Margaret Mead are coming close (e. g. , Freeman, 1983).

55. For the assault on Gould, see *The New York Review of Books*, June 12, June 26, and August 14, 1997, and Dawkins (1998), Alcock (1998), and Wright (1999).

56. Watson (1925).

57. Lorenz (1952, p. 146).

58. Manning (1996).

59. Bischof (1991), for example, has written a Lorenz "psychogram" that reads like a psychoanalysis of the author's own father complex vis-à-vis the intimidating Austrian ethologist.

60. The Russian manuscript (1944–1948) was published as *Die Naturwissenschaft vom Menschen* (Munich: Piper, 1992). The original manuscript consisted of 750 pages written in diluted ink with quills and steel pens. Lorenz often traded his meager food rations for writing materials. At the worst moment during his stay in Russia, he weighed 55 kg.

61. For sources of the quotations provided here, and documentation of Lorenz's activities before and during World War II, see Deichmann (1996).

62. Kalikow (1980).

63. In a televised interview in 1981, Lorenz explained: "That they meant murder when they said 'elimination' or 'selection' was something I really did not believe at the time. This is how naïve, how stupid, how gullible—call it what you will—I was back then" (Deichmann, 1996).

64. Lorenz was mentioned as *ehrenamtliche* as opposed to *hauptamtliche Mitarbeiter.*

65. The eugenics movement was founded by Francis Galton, a cousin of Charles Darwin, and further developed by Karl Pearson in the beginning of the twentieth century at the University College, London. Pearson felt that "superior" races must supplant "inferior" ones. Not surprisingly, the term "eugenics" became common fare in Hitler's propaganda. Further see Gould (1981).

66. Defenders of Lorenz note the absence of blatant racism, saying that he had too many Jewish friends to be anti-Semitic, and that at the most he opposed racial mixing (Bischof, 1991). Pessimism about racial mixing continues, albeit in a different form, in the writings of a well-known Lorenz student: Eibl-Eibesfeldt (1994) has argued that human xenophobic tendencies prevent a fully integrated multiethnic society.

67. Quoted in Deichmann (1996, p. 179).

68. Robert Hinde, personal communication.

69. Lorenz (1985).

70. From a letter to Margaret Nice, reproduced in full in Deichmann (1996). Dated June 23, 1945, this letter was written immediately after the liberation of the Netherlands, at a time when Lorenz was still imprisoned.

71. For example, Maarten 't Hart, a Dutch ethologist and novelist, acidly commented in a review of a book by Lorenz: "I still see before me (it was clearly visible on television) how, during the award ceremony of the Nobel Prize, he came forward with all those medals and orders pinned to his chest. How many distinctions from between 1933 and 1942 did these include? And if he had left those off—which I hope—why did he need to wear the others?" (*NRC Handelsblad*, March 14, 1989).

72. For a recent example, see the last chapter of Ridley's (1996) *The Origins of Virtue*. Ridley explores the common ground between gene-centric evolutionary biology and a conservative political agenda, quoting Margaret Thatcher: "There is no such thing as society. There are individual men and women, and there are families."

73. Liessmann (1996).

74. Wilson (1995).

75. L. B. Halstead (1984), *Kinji Imanishi: The View from the Mountain Top*. Unpublished English manuscript in the Kyoto University Library, later published in Japanese.

76. As noted by Inoue and Anderson (1988), Halstead stayed a little too long because "in the book he passes from fresh first impressions to oversimplistic theories about the nature of Japan. His final view, that Imanishi's theory has been popular because it offers a dream world of harmony to the cruel reality of modern Japan, is naïve."

77. The book was published in a long tradition of translated works that confirm to the Japanese how hard it is for outsiders to understand them. The book probably amused as much as it insulted.

78. Halstead (1985).

79. Yoshimi (1998).

80. This is not to say that there exists consensus among ecologists about the role of competition between species in population dynamics and speciation (e. g. , Sinclair, 1986).

81. See Asquith (1991) and Sakura (1998).

82. In 1958, the Japan Monkey Center sent two scientists to Africa, Imanishi and his student, Itani. They saw gorillas at the Virunga Mountains, but only heard chimpanzees in Cameroon. On the second trip, Itani went on his own all across Africa, where—as was typical for his combined interest in primatology and ethnography—he looked for both apes and human pygmies. Ignoring Leakey's prohibition, he also stopped by Jane Goodall's camp, which she had started only two months earlier. In 1961, the first Japanese studies of chimpanzee behavior began at various field sites in Tanzania, resulting in the establishment of the Mahala Mountains project in 1965. This influential project is still in operation today, headed by Toshisada Nishida, one of Itani's most prominent students of ape behavior. Another well-known Itani student is Takayoshi Kano, who, in 1973, set up the only long-term field project on the elusive bonobo, in the Democratic Republic of Congo. The use of stone tools by chimpanzees was first described by Yukimaru Sugiyama (see Chapter 7, "The Nutcracker Suite").

83. Goodall (1990) reveals a glimpse of the initial Western resistance to the application of ethnographic methods to animals, relating how one respected scientist told her that even if all individuals were different it would still be best to sweep this fact under the carpet. Editors of journals objected to her grammar: "The editorial comments on the first paper that I wrote for publication demanded that every *he* or *she* be replaced with *it*, and every *who* be replaced with *which.* "

84. Inoue and Anderson (1988).

85. Quoted from an unpublished lecture by Mariko Hiraiwa-Hasegawa at the Department of Biology, Princeton University, February 26, 1992: "Sociobiology and Japanese Primatology: A Case of Struggle for Survival in a Conformist Society. "

86. I closely followed the response to sociobiology in the Netherlands, and even edited a Dutch volume about it. Whereas the popular media and social sciences showed strong resistance, its reception in the scientific community was uneventful. After all, Dutch ethology had produced Niko Tinbergen, who exported the evolutionary study of animal behavior to Oxford, where it became a cornerstone of the British contribution to the so-

ciobiological revolution. Thus, Dutch ethologists recognized continuity between their own tradition and the "new" synthesis.

3. Bonobos and Fig Leaves
Primate Hippies in a Puritan Landscape

87. In one such study, on our favorite chimpanzee group at the Yerkes Field Station, we had an average of only one or two spontaneous aggressive conflicts per day. This meant that the student collecting the data, Xin Wang, would sit on an observation tower in the hot Georgia sun for entire days, and collect information on anywhere from zero to perhaps five fights a day. Our interest was to see what happened afterward—did the chimpanzees reconcile with a kiss or grooming, or not?—but such a study requires at least a couple of hunderd instances. When the data are summarized in a nice, colorful graph, no one realizes how much time went into obtaining them.

88. I am referring to the Monica Lewinsky affair of 1998–99, during which U. S. President Clinton's political adversaries were perplexed by the lack of moral outrage in their country. This entire scandal, including the media feeding frenzy on sexual innuendo, illustrated what Charles Dickens termed the "attraction of repulsion. " That is, the only way sex can be openly debated in a puritanical society is within the context of concern and disapproval.

89. The male body is subject to even greater taboos than the female body. In the United States one will not see the tiny swimming trunks of many European men—which reveal and even accentuate what's inside—nor unclad male statues, such as "Manneke Pis," the proud symbol of Brussels. Even a life-sized sculpture of an elephant, donated by African governments to the United Nations, had to be surrounded in a hurry by potted plants and shrubs so as to block a side view of an animal that was so undeniably male. It had never crossed the mind of the Bulgarian-born artist, Mihail, that an anatomically correct animal could stir up controversy in New York. Given that the elephant was to represent environmental awareness, he commented: "This is exactly the problem between people and wildlife, people cannot face nature. " From CNN on the Internet: "U. N. Elephant Statue Draws Guffaws for Being Too Long on Realism" (November 18, 1998).

90. Ehrenreich (1999).

91. The editorial note appeared in *Time* of March 29, 1999. Protest against the androgynous photos was published in *Time* of April 5, 1999. A

physician from the country of Peter Paul Rubens called the defeminized bodies in the magazine "anatomical heresy. "

92. Diamond (1990).

93. Animals may show homosexual behavior but are generally not homosexual in the sense of having an exclusive or predominant orientation toward same-sex partners.

94. Parish (1993).

95. Kano (1998).

96. Bagemihl (1999, p. 117).

97. Stanford (1998). In the most detailed comparison to date, I observed the sexual behavior of bonobos at the San Diego Zoo and that of chimpanzees at the Field Station of the Yerkes Primate Center in Atlanta. The average adult bonobo initiated sex once every 65 minutes, in contrast to once every 6 hours for the average chimpanzee under similar conditions (de Waal, 1995). Even if these rates show that bonobos have sex far more often than chimpanzees, it is also evident that they are not doing it all the time. I must admit, though, that once when I tried to make this point by explaining that bonobos had sex only once every hour, an editor wrote in the margin of my manuscript that this sounded very much like all the time to her.

98. Kuroda (1984).

99. One sex therapist in California, Susan Block, has selected the bonobo ("the horniest chimps on earth") as her favorite animal, advertising her business on her Web site with: "whether you want to help save the bonobos or just save your own sex life, I'll be here for you. "

100. Parish and de Waal (2000).

4. Animal Art
Would You Hang a Congo on the Wall?

101. Introduction by Morris to Lenain (1997).

102. Deacon (1999).

103. Leakey and Lewin (1992).

104. Gilliard (1969).

105. Miller (2000).

106. Quoted in Hildebrand (1999).

107. Porter and Neuringer (1984) and Watanabe and Nemoto (1998).

108. Marler and Tamura (1964). This is not to say that there is no biological effect on how a bird sings. Many birds learn the song of their own species more accurately than that of another species, and if exposed to both,

tend to learn the former, which means that learning is biased toward their own species.

109. L. F. Baptista, at the Annual Meeting of the American Association for the Advancement of Science, Washington, D. C. , 2000. The speaker died a few months later.

110. My translation from the first part of Mozart's poem: "Hier ruht ein lieber Narr, ein Vogel Staar. Noch in den besten Jahren, mußt er erfahren des Todes bittern Schmerz. "

111. Liner notes accompanying A Musical Joke by W. A. Mozart, Deutsche Grammophon 400 065–2.

112. West and King (1990, p. 112).

113. Watanabe et al. (1995). During my visit to his lab, I asked Watanabe why he tested his birds on Western painters and composers—why not Japanese? He answered that the reviewers of international journals are impressed only by birds discriminating among art that they know, or have heard about.

114. Busch and Silver (1994).

115. Huxley (1942).

116. Schiller (1951). Alpha was the first-born chimpanzee of the Yerkes colony at Orange Park in Florida, the precursor of the Yerkes Regional Primate Research Center, where I now work, which is part of Emory University in Atlanta, Georgia.

117. Lenain (1997).

118. The book's translation out of French erroneously uses the term "monkey" as synonymous with "ape," perhaps because the distinction is less clear in French (singe and grand singe, respectively). In contrast to monkeys, apes belong to the hominoids, a small, distinct primate family consisting only of bonobos, orangutans, gorillas, chimpanzees, gibbons, and humans.

119. Lenain (1997) notes how disruption of order requires that one recognizes order, hence has a sense of it. So he doesn't see his position as totally at odds with Morris's. However, where Morris postulates a positive tendency to create something, Lenain sees the ape as trying to get rid of something else. The height of either confusion or brilliance is reached when Lenain states: "It is perfectly conceivable that a kind of 'sense of disorder' reigns in the image field so straightforwardly that it may manifest itself mainly in the guise of a sense of order. "

120. Apes find painting pleasurable and self-reinforcing. They do it enthusiastically without any outside reward. To test this, Morris reinforced one chimpanzee with tidbits for any artistic expression. The result was a dra-

matic loss of interest: the ape worked as quickly as possible, only to hold out
a hand for the reward (Morris, 1962).

121. Levy (1961).

5. Predicting Mount Fuji, and a Visit to Koshima, Where the Monkeys Salt Their Potatoes

122. This is a paraphrasing of Imanishi's insightful discussion of culture
(Itani and Nishimura, 1973), which now seems rather unremarkable, but
was published in 1952, at about the time that Western scientists were still
engaged in a polarized debate about whether behavior depended on learning or instinct.

123. Kurland (1977).

124. Sugiyama (1967).

125. Sommer (1994).

126. In some species, females seem to make paternity a confused issue—
for example, by having sexual encounters with males even when they are
not fertile. If males are unable to exclude their own offspring from infanticide, their strategy becomes counterproductive according to theories proposed by Hrdy (1979).

127. Mayr (1997). My own experience in this regard was the discovery of
reconciliation. That cooperative animals need to repair their relationships
after fights sounds logical enough, but reconciliation behavior was not predicted or even remotely considered by evolutionary biologists, who traditionally have shown far more interest in win-lose than win-win arrangements.
My initial hunches, based on seeing chimpanzees kiss and embrace after a
fight, are now supported by studies of a host of species (Aureli and de Waal,
2000; de Waal, 2000).

128. Wolpert (1992).

129. Nishida (1990).

130. Asquith (1986).

131. Asquith (1989, pp. 136–137).

132. Imanishi (1952).

133. *Ko* means "happy" in Japanese, and *shima* means "island." Koshima
island is therefore redundant. Other famous monkey sites in Japan also have
built-in habitat descriptions, such as Yakushima, Takasakiyama, Arashiyama,
Ryozenyama, Jigokudani, and Katsuyama (*yama* means mountain, *dani*
means valley).

134. My enthusiastic interpreter was Satsuki Kuroki. On my visit to the island, I was accompanied by Kunio Watanabe, a scientist from the Primate Research Institute in Inuyama, who has worked at Koshima for many years. See Watanabe (1994).

135. Kawamura wrote the pioneering papers on the Koshima monkeys, laying out the argument for cultural propagation. He also ascribed intergroup differences to differing traditions, noting how some monkey troops eat eggs, while others do not, or how paternal care is restricted to some troops. To avoid battles with Japanese skeptics, he used the term "subculture" rather than "culture. " Most of his studies appeared in Japanese only (Itani and Nishimura, 1973).

136. In the first five years, which Kawai (1965) called the period of "individual propagation," 15 out of 19 monkeys between the ages of two and seven years acquired the behavior, but only 2 out of 11 adults. During the following period of "precultural propagation," almost all infants born to potato washing mothers learned the habit. Ten years after Imo's discovery, ninety-seven percent of the monkeys under the age of twelve years showed the habit.

137. Noso died a few months after my visit, and the beta male, Kemushi, took his place.

138. Watanabe (1989).

139. Keyes (1982, pp. 14-17).

140. At five other Japanese provisioning sites, monkeys developed potato washing. At these sites, the behavior never spread, though; it remained restricted to a few isolated individuals. These observations do show that Imo did not exactly discover the monkey equivalent of the wheel: the cleaning of food in water develops quite readily (Visalberghi and Fragaszy, 1990a).

141. Keyes (1982). For a thorough debunking of this piece of pseudoscience, see Amudson (1985).

142. Galef (1990).

143. Steven Green visited Japan in 1968 and 1969, and attended potato feedings at Koshima (Green, 1975).

144. Galef (1990), an investigator of laboratory rats, failed to consider this constraint on food provisioning. His remarkable free association about what *may* have occurred at Koshima, forty years after the fact, has never before been critically examined. The scientists involved didn't know how to politely respond to harsh criticism concerning an issue imitation—that they themselves had never emphasized.

145. Recently, the first report appeared of Japanese monkeys developing a washing habit without any human influence, involving a natural food. Monkeys at Katsuyama were seen to pull grass roots one by one from the ground, then carry a pile of them to a nearby river to wash off the dirt before eating them. The habit spread to eleven adult females, six of whom belonged to the same matriline (Nakamichi et al. , 1998).

146. Galef (1990).

147. More relevant than the speed of learning is the *shape* of the learning curve in the population, which should be linear in the case of individual learning but accelerating over time in the case of social learning. A detailed analysis by Lefebvre (1995) of learning rates contradicts claims of individual learning, and lends support to the cultural learning model.

6. The Last Rubicon
Can Other Animals Have Culture?

148. In an exceptionally open-minded chapter on the background of culture, American anthropologist Ralph Linton (1936) warned that, although human culture is unique, some of the underlying abilities can be traced to "the animal level. " In line with the argument presently pursued, he noted that the exact learning process is irrelevant, "for the vital thing in the transmission of learned behavior has been the ability of each generation to take over the habits of the one preceding it. "

149. For more on Imanishi and his school of primatology, see "The Fate of Gurus" (Chapter 2) and "Predicting Mount Fuji" (Chapter 5).

150. Shweder (1991) discusses "Post-Nietzchean" anthropology and how it treats cultural practices as arbitrary and imaginary, resulting in a degrading of custom and tradition as well as a rupture with the natural world.

151. Tylor (1871).

152. White (1959).

153. In its negative form (what culture is *not*) the "how" question is critical, however. Cultural propagation is defined as nongenetic, as in one of the earliest inclusive definitions by Bonner (1980): "By culture I mean the transfer of information by behavioral means, most particularly by the process of teaching and learning. It is used in a sense that contrasts with the transmission of genetic information passed by direct inheritance of genes from one generation to the next. " For my own definition, see the "Prologue. "

154. Ladygina-Kohts (in press).

155. Russon (1996, p. 166).

156. Tomasello, Savage-Rumbaugh, and Kruger (1993). For the concept of enculturation, see Tomasello and Call (1997): "The most plausible hypothesis at present is that human-raised apes understand the intentions of others in ways that their wild conspecifics do not. " The little problem that I see in this conclusion is with the word "others," which ought to be replaced by "humans. "

157. Despite my criticism, the ape-human imitation paradigm recently worked in an ingenious experiment by Whiten (1998). One such positive outcome obviously puts question marks behind the negative results of similar experiments in the literature.

158. Humphrey (1976) offers a definition of sympathy: "By sympathy I mean a tendency on the part of one social partner to identify himself with the other and so make the other's goals to some extent his own. "

159. Darwin (1871).

160. See Byrne and Russon (1998) and its twenty-eight commentaries.

161. Köhler (1925, p. 280).

162. Matsuzawa (1994) and Inoue-Nakamura and Matsuzawa (1997).

163. Galef (1990).

164. Huffman (1996).

165. As a conformist form of learning, BIOL fits current evolutionary thinking about the origins of culture (Henrich and Boyd, 1998).

166. As an illustration of the polarized nature of the ongoing debate, this paper appeared under the uncompromising title "Why Animals Have Neither Culture nor History" (Premack and Premack, 1994).

167. That Sarah readily imitates people (which is in line with similar findings on language-trained apes) fully agrees with the BIOL model, according to which social learning is facilitated by emotional closeness to and identification with the model. The difference from the enculturation hypothesis is that BIOL attributes Sarah's imitation to her interest in and familiarity with human companions, whereas the enculturation hypothesis makes the far less parsimonious assumption of altered mental faculties.

168. Kroeber (1923, p. 104).

169. Gould (1999), Whiten et al. (1999), and de Waal (1999).

170. Montagu (1968).

171. McGrew (1992) protests against the half-accept, half-reject approach that dominated the early literature on animal culture, when authors would place "culture" between quotation marks, or add prefixes, as in

"proto-culture," "pre-culture," or "sub-culture. " He notes that there is no justification for these neologisms and that "the coining of new terms is no substitute for explicit reasoning. " In the same influential book, McGrew pays much attention to Kroeber's views and shows how they relate to the question of chimpanzee culture.

172. Kroeber (1928).

7. The Nutcracker Suite
Reliance on Culture in Nature

173. Sept and Brooks (1994).

174. Beatty (1951, p. 118).

175. Sugiyama and Koman (1979).

176. Boesch and Boesch-Achermann (1991, p. 53). In this quote from *Natural History,* I have translated medieval measures into metric.

177. For sex differences in tool use, see McGrew (1979) and Boesch and Boesch (1984).

178. The claim that tool making separates us from other life forms was made most forcefully by Kenneth Oakley (1957) in *Man the Tool-Maker.* Oakley was aware of Köhler's observations that chimpanzees sometimes fashion tools, for example, by fitting a stick into a bamboo tube so as to make it longer. Oakley refused to count what the apes did as tool manufacture, however, since it was done in reaction to a given situation—a banana dangling outside the cage—rather than in anticipation of an imagined future.

179. In contrast, the most quoted definition of tool use among students of animal behavior is the one by Beck (1980): "the external deployment of an unattached environmental object to alter more efficiently the form, position, or condition of another object. "

180. Allen (1997, p. 48).

181. Due to articles by nonprimatologists (e. g. , Mann, 1972), the idea has become widespread that chimpanzee behavioral diversity has a negligible impact on subsistence. This unproven supposition has become another weapon against the "culture" label. However, as is evident from examples in this chapter, a great deal of chimpanzee culture concerns food collection, including foods that are inaccessible without complex learned techniques. Further, one also wonders where the requirement that culture is crucial for survival leaves *human* culture: many human cultural variants have little or nothing to do with survival.

182. Günther and Boesch (1993).

183. Yamakoshi (1998).

184. McGrew and Tutin (1978). Handclasp grooming is now known of several other chimpanzee communities in the field, such as one in Uganda observed in the 1980s by Ghiglieri (1988).

185. De Waal and Seres (1997).

186. De Waal (1989a).

187. De Waal (1989a).

188. Marshall et al. (1999).

189. Hirata et al. (1998).

190. Alp (1997).

191. Most of the evidence has been summarized by Huffman (1997). At the Primate Research Institute of Kyoto University, Huffman is currently exposing chimpanzees to bristly, rough leaves similar to *Aspilla*. Their first response is avoidance, but the aversion is overcome by apes who have watched another put leaves into its mouth.

192. Hinde and Fisher (1951).

193. Aisner and Terkel (1992).

194. Whiten (1998).

195. Van Schaik et al. (1999). For inhibition in a less tolerant primate, see Drea and Wallen (1999).

196. Tanaka (1995).

197. Nakamura et al. (in press). Empathy and perspective-taking in apes are discussed in Chapter 2 of de Waal (1996a).

198. Boesch (1991).

199. Boesch and Tomasello (1998).

200. The reduction of culture to discrete self-propagating units is rather old. Recent genetically inspired terms for these units are "meme" by Dawkins (1976) and "culturgen" by Lumsden and Wilson (1981). Most scientists have given these ideas little thought, considering them "cocktail-party science" (*Time*, April 19, 1999). For a critical yet supportive comparison of genes with memes, see Wimsatt (1999).

201. See Lumsden and Wilson (1981), Boyd and Richerson (1985), and Durham (1991).

202. The term "cultural biology" was already proposed in the 1950s by Imanishi when he sought to foster contact between anthropologists and zoologists (T. Nishida, personal communication).

203. Payne (1998).

204. Ottoni and Mannu (in press) have documented more than one hundred cracking sites in the park, which is near São Paulo. They also observed the monkeys' behavior and found that, as in chimpanzees, the young are less efficient crackers than the adults. Despite increasing field research on these primates, this is the first wild capuchin group for which stone-tool use has ever been reported.

205. The 1998 Prix Jean-Marie Delwart awarded by the Royal Academy of Sciences of Belgium.

206. Wrangham et al. (1994) and Whiten et al. (1999). For further information see http://chimp. st-and. ac. uk/cultures/. Apart from these books and articles, a recent French edited volume posed the question whether culture is natural (Ducros et al. , 1998).

207. Exclusion of ecological explanations for a single behavioral variant, such as nut cracking, has been undertaken by McGrew et al. (1997).

208. Whitehead (1998).

8. Cultural Naturals: Tea and Tibetan Macaques

209. Corbey (1997).

210. Montagu (1968).

211. In the late fall, when they are at their heaviest, adult males weigh on average 19. 5 kg and adult females 16. 8 kg (Zhao, 1996).

212. The term "sexual harassment" was in use in the primate literature long before it became a public issue in Western society. The term refers to the disturbance of mating pairs by bystanders, which may jump on top of them, pull their hair, and so forth.

213. The most detailed studies of bridging in Tibetan macaques are those by a Japanese researcher, Hideshi Ogawa. Making his observations at Huang Shan, Ogawa found that adult males bridge more with male than with female infants. When approaching a dominant male, they often bring the particular infant that this male prefers (i. e. , the infant he associates with most). Bridging also occurs between the sexes, and among females. For these encounters, the offspring of the approached female is often used. The selection of infants thus suggests knowledge of which infant is attractive to the bridging partner (Ogawa, 1995).

214. Chapais (1988).

215. De Waal (1996b).

216. This is a realistic example. Under crowded conditions, rhesus monkeys groom and reconcile more outside their own matrilines. Possibly, they

are trying to prevent strained relations between matrilines. If the tendency to seek contact outside the family is transmitted from mother to offspring, crowding can be said to induce fundamental changes in rhesus social culture (Call et al., 1996).

217. I prefer the term "cultural natural" to "cultural universal," which is found in the anthropological literature. The latter term maintains a separation between culture and nature, as if patterns found in all cultures are likely to be independent of human biology.

218. Kummer (1995, p. 12?).

219. De Waal and Johanowicz (1993).

220. See also Chapter 11 ("Down with Dualism").

221. Small (1988). See also Hrdy (1999).

9. Apes with Self-Esteem
Abraham Maslow and the
Taboo on Power

222. Maslow (1936).

223. See also Noso on Koshima Island, who was in a similar "propped up" position as Spickles (Chapter 5, "Predicting Mount Fuji").

224. Cullen (1997).

225. See Chapter 1 ("The Whole Animal").

226. De Waal (1982). For comparable political strategies in wild chimpanzees, see especially Nishida and Hosaka (1996).

227. Woodward and Bernstein (1976).

228. The incident in Arnhem has been described in detail in de Waal (1986). For the parallel incident at Gombe, see Goodall (1992).

229. As argued in *Chimpanzee Politics*, alliances among male chimpanzees are not based on friendship or personal liking; they are entirely strategic. Hence, with the disappearance of their common rival, Nikkie, the reason for the Yeroen-Dandy alliance evaporated.

230. Mulder (1979).

231. Barkow (1975).

232. For systematic research on human reactions to the nonverbal displays of political leaders, see Masters (1989). In relation to turning off the sound of my TV, the following observation by neurologist Oliver Sacks (1985) is of interest. He describes a group of patients in the aphasia ward convulsed with laughter during a televised speech by their nation's president (characterized by Sacks as "the old Charmer" and "the Actor"). Incapable of

understanding words as such, aphasia patients nevertheless follow much of what is being said by means of the facial expressions and other body language accompanying speech. They are so good at processing nonverbal clues that they have the reputation that one cannot lie to them: they see right through lies. Sacks concluded about the president's speech, which seemed perfectly normal to the nonpatients present, that it "so cunningly combined deceptive word-use with deceptive tone, that only the brain-damaged remained intact, undeceived. "

233. Even the most egalitarian human societies don't manage to get rid of the dominance drive. Instead, Boehm (1999) speaks of "leveling mechanisms," meaning that status differences are actively suppressed. In these societies, men trying to wield power risk ridicule and ultimately resentment.

10. Survival of the Kindest
Of Selfish Genes and Unselfish Dogs

234. In the 1975 Christmas season, millions of Americans spent five dollars each to purchase ordinary rocks as pets. The rocks were sold in boxes with air holes and came with a manual explaining how to train the rock to roll over, to play dead, and to protect its owner. Tamagotchi is a popular Japanese electronic gadget that mimics a chick. It eats, sleeps, defecates, gets cranky, and beeps for attention. If the owner does not take care of it, Tamagotchi dies.

235. Wrangham and Peterson (1996). Playing with "dolls" is not unusual in nonhuman primates. I have seen young chimpanzees in captivity act the same as Kakama with a piece of cloth or a broom. A wild mountain gorilla was seen to pull up a mass of soft moss, which she carried and held like an infant under her chest, cuddling and "nursing" it (Byrne, 1995).

236. Darwin (1859).

237. Quoted from an interview by Roes (1998).

238. Also, let us not forget that many people volunteer to adopt children—some even kidnap newborns from the maternity ward—following urges that evidently transcend genetic self-interest.

239. See Chapter 1 ("The Whole Animal").

240. Reported in *The Jerusalem Post*, July 26, 1996.

241. De Waal and Aureli (1996).

242. Ladygina-Kohts (in press).

243. Kunz and Allgaier (1994).

244. Jewell (1997).
245. Whittemore and Hebard (1995).
246. Whittemore and Hebard (1995).

11. Down with Dualism!
Two Millennia of Debate
About Human Goodness

247. Westermarck (1912).
248. De Waal and Luttrell (1988) and Aureli et al. (1992).
249. For recent debate about evolutionary ethics, see the *Journal of Consciousness Studies*, vol. 7 (1–2), edited by L. D. Katz (2000).
250. Wolf (1995). Others before him studied marriages in Israeli kibbutzim and found that children do not have sexual intercourse, let alone marry unrelated children of the opposite sex with whom they have grown up in the same peer group (reviewed by Wolf, 1995).
251. Tokuda (1961–62).
252. Huxley (1894).
253. Desmond (1994).
254. Freud (1913, 1930).
255. De Waal (1996a). See also Flack and de Waal (2000).
256. Williams quoted in Roes (1998), Dawkins in *Times Literary Supplement* (November 29, 1996), and Dawkins in another interview by Roes (1997). The profound irony, of course, is that contrary to Dawkins's warning against a Darwinian world, such a world is eminently more livable than a Huxleyan one, which is devoid of natural moral tendencies. Dawkins seems almost a reincarnation of Huxley in terms of both combativeness (c. g. , Dawkins, 1998) and his departure from Darwinism. Such notions as that we are survival machines, that we are born selfish and need to be taught kindness, and especially that morality and biology are miles apart were alien to Darwin yet typical of Huxley. Darwin never looked at any life form as a machine. He had a Lorenz-like rapport with animals and didn't shy away from attributing intentions and emotions to them. Crist (1999) discusses at length Darwin's anthropomorphism, which has irritated some scholars, but confirms that those with an integrated view of nature don't necessarily have a problem with it (see also Chapter 1, "The Whole Animal"). Given their differences of opinion, Darwin couldn't resist referring, in his final letter to Huxley, to the latter's depiction of all living things (including humans) as

machines: "I wish to God there were more automata in the world like you. " (Cited in Crist, 1999).

257. In view of their cynical positions, the titles of the books by Wright (*The Moral Animal*) and Ridley (*The Origins of Virtue*) don't exactly cover their message (Wright, 1994; Ridley, 1996).

258. Wright (1994).

259. Sober and Wilson (1998) write about this accusation: "We feel we should address a criticism that is often leveled at advocates of altruism in psychology and group selection in biology. It is frequently said that people endorse such hypotheses because they *want* the world to be a friendly and hospitable place. The defenders of egoism and individualism who advance this criticism thereby pay themselves a compliment; they pat themselves on the back for staring reality squarely in the face. Egoists and individualists are objective, they suggest, whereas proponents of altruism and group selection are trapped by a comforting illusion. "

260. Ideas about the subconscious and its evolutionary *raison d'être* have been around since Badcock (1986) and Alexander (1987). The first explicitly sought to provide Freudian-Darwinian solutions to the "problem" of altruism.

261. Mayr (1997).

262. Damasio (1994).

263. Aureli and de Waal (2000).

264. Westermarck lists moral approval as a kind of retributive kindly emotion, hence as a component of reciprocal altruism. These views antedate discussions about "indirect reciprocity" and reputation building in the modern literature on evolutionary ethics (e. g. , Alexander, 1987).

265. Smith (1759).

266. These reflections by Westermarck parallel Smith's (1759) idea of an "impartial spectator. "

267. Darwin (1871).

268. Reviewed by Preston and de Waal (in press).

269. Darwin (1871).

270. This makes Mencius a contemporary of Aristotle—born 384 B.C. in Greece—the first and foremost Western philosopher to root morality in human biology (Arnhart, 1998).

271. All quotations are from Mencius (372–289 B.C.), *The Works of Mencius.*

272. See Chapter 10 ("Survival of the Kindest"), which also contains the full quotation from Smith.

273. Killen and de Waal (2000).

Epilogue

274. The puffer fish has an extremely toxic liver, which, if not appropriately removed, causes paralysis and certain death (not surprisingly, preparing this fish for *fugu* requires a special license in Japan). The risky consumption of this delicacy is comparable to the bitter-pith chewing of wild chimpanzees, in which the apes seem to have learned to avoid the toxic parts of a particular plant (Huffman, 1997).

Bibliography

Aisner, R., and Terkel, J. (1992). Ontogeny of pine-cone opening behaviour in the black rat (*Rattus rattus*). *Animal Behaviour* 44: 327–336.

Alcock, J. (1998). Unpunctuated equilibrium in the *Natural History* essays of Stephen Jay Gould. *Evolution and Human Behavior* 19: 321–336.

Alexander, R. A. (1987). *The Biology of Moral Systems*. New York. Aldine de Gruyter

Allen, B. (1997). The chimpanzee's tool. *Common Knowledge* 6: 34–51.

Alp, R. (1997). "Stepping-sticks" and "seat-sticks": New types of tools used by wild chimpanzees (*Pan troglodytes*) in Sierra Leone. *American Journal of Primatology* 41: 45–52.

Amudson, R. (1985). The hundredth monkey phenomenon. *The Skeptical Enquirer* 9: 348–356.

Arnhart, L. (1998). *Darwinian Natural Right: The Biological Ethics of Human Nature*. Albany, NY: SUNY Press.

Asquith, P. J. (1986). Anthropomorphism and the Japanese and Western traditions in primatology. In J. G. Else and P. C. Lee (eds.), *Primate Ontogeny, Cognition, and Social Behavior*, pp. 61–71. Cambridge: Cambridge University Press.

Asquith, P. J. (1989). Provisioning and the study of free-ranging primates: History, effects, and prospects. *Yearbook of Physical Anthropology* 32: 129–158.

Asquith, P. J. (1991). Primate research groups in Japan: Orientations and East-West differences. In L. Fedigan, and P. Asquith (eds.), *The Monkeys of Arashiyama: Thirty-five Years of Research in Japan and the West*, pp. 81–98. Albany, NY: SUNY Press.

Atkins, K. A. (1996). A bat without qualities? In M. Bekoff and D. Jamieson (eds.), *Readings in Animal Cognition*, pp. 345–358. Cambridge, MA: MIT Press.

Aureli, F., Cozzolino, R., Cordischi, C., and Scucchi, S. (1992). Kin-oriented redirection among Japanese macaques: An expression of a revenge system? *Animal Behaviour* 44: 283–291.

Aureli, F., and de Waal, F. B. M. (2000). *Natural Conflict Resolution.* Berkeley: University of California Press.

Austin, W. A. (1974). *The First Fifty Years: An Informal History of the Detroit Zoological Park and the Detroit Zoological Society.* Detroit: The Detroit Zoological Society.

Badcock, C. R. (1986). *The Problem of Altruism: Freudian-Darwinian Solutions.* Oxford: Blackwell.

Bagemihl, B. (1999). *Biological Exuberance: Animal Homosexuality and Natural Diversity.* New York: St. Martin's.

Bailey, M. B. (1986). Every animal is the smartest: Intelligence and the ecological niche. In R. Hoage and L. Goldman (eds.), *Animal Intelligence,* pp. 105–113. Washington, D. C. : Smithsonian Institution Press.

Balda, R. P., and Kamil, A. C. (1989). A comparative study of cache recovery by three corvid species. *Animal Behaviour* 38: 486–495.

Barkow, J. H. (1975). Prestige and culture: A biosocial interpretation. *Current Anthropology* 16: 553–572.

Batson, C. D., Early, S., and Salvarani, G. (1990). Perspective taking: Imagining how another feels versus imagining how you would feel. *Personality and Social Psychology Bulletin* 23: 751–758.

Beach, F. A. (1950). The snark was a boojum. *American Psychologist* 5: 115–124.

Beatty, H. (1951). A note on the behavior of the chimpanzee. *Journal of Mammalogy* 32: 118.

Beck, B. B. (1980). *Animal Tool Behavior: The Use and Manufacture of Tools by Animals.* New York: Garland.

Bischof, N. (1991). *Gescheiter als alle die Laffen.* Hamburg: Rasch & Röhring.

Boehm, C. (1999). *Hierarchy in the Forest: The Evolution of Egalitarian Behavior.* Cambridge, MA: Harvard University Press.

Boesch, C. (1991). Teaching in wild chimpanzees. *Animal Behaviour* 41: 530–32.

Boesch, C., and Boesch, H. (1983). Optimization of nut-cracking with natural hammers by wild chimpanzees. *Behaviour* 83: 265–286.

Boesch, C., and Boesch, H. (1984). Possible causes of sex differences in the use of natural hammers by wild chimpanzees. *Journal of Human Evolution* 13: 415–440.

Bibliography 391

Boesch, C., and Boesch-Ackermann, H. (1991). Dim forest, bright chimps. *Natural History* 9/91: 50–56.

Boesch, C., and Tomasello, M. (1998). Chimpanzee and human cultures. *Current Anthropology* 39: 591–614.

Bonner, J. T. (1980). *The Evolution of Culture in Animals*. Princeton, NJ: Princeton University Press.

Boyd, R., and Richerson, P. J. (1985). *Culture and the Evolutionary Process*. Chicago: University of Chicago Press.

Budiansky, S. (1998). *If a Lion Could Talk*. New York: Free Press.

Burghardt, G. M. (1985). Animal awareness: Current perceptions and historical perspective. *American Psychologist* 40: 905–919.

Busch, H., and Silver, B. (1994). *Why Cats Paint: A Theory of Feline Aesthetics*. Berkeley, CA: Ten Speed.

Byrne, R. W. (1995). *The Thinking Ape*. Oxford: Oxford University Press.

Byrne, R. W., and Russon, A. E. (1998). Learning by imitation: A hierarchical approach. *Behavioral and Brain Sciences* 21: 667–721.

Call, J., Judge, P. G., and de Waal, F. B. M. (1996). Influence of kinship and spatial density on reconciliation and grooming in rhesus monkeys. *American Journal of Primatology* 39: 35–45.

Cenami Spada, E. (1997). Amorphism, mechanomorphism, and anthropomorphism. In R. W. Mitchell, N. S. Thompson, and H. L. Miles (eds.), *Anthropomorphism, Anecdotes, and Animals*, pp. 37–49. Albany, NY: SUNY Press.

Chapais, B. (1988). Rank maintenance in female Japanese macaques: Experimental evidence for social dependency. *Behaviour* 104: 41–59.

Chelser, P. (1969). Maternal influence in learning by observation in kittens. *Science* 166: 901–903.

Cheney, D. L., and Seyfarth, R. M. (1990). *How Monkeys See the World*. Chicago: University of Chicago Press.

Corbey, R. (November 8, 1997). Beschaving is meer dan mes en vork. *NRC Handelsblad*.

Crist, E. (1999). *Images of Animals: Anthropomorphism and Animal Mind*. Philadelphia: Temple University Press.

Cullen, D. (1997). Maslow, monkeys, and motivation theory. *Organization* 4: 355–373.

Curio, E. (1978). Cultural transmission of enemy recognition: One function of mobbing. *Science* 202: 899–901.

Custance, D. M., Whiten, A., and Bard, K. A. (1995). Can young chimpanzees imitate arbitrary actions? Hayes and Hayes (1952) revisited. *Behaviour* 132: 839–858.

Damasio, A. R. (1994). *Descartes' Error: Emotion, Reason, and the Human Brain*. New York: Putnam.

Damasio, A. R. (1999). *The Feeling of What Happens*. New York: Harcourt.

Darwin, C. (1964 [1859]). *On the Origin of Species*. Cambridge, MA: Harvard University Press.

Darwin, C. (1981 [1871]). *The Descent of Man, and Selection in Relation to Sex*. Princeton, NJ: Princeton University Press.

Darwin, C. (1998 [1872]). *The Expression of the Emotions in Man and Animals*. Third Edition. New York: Oxford University Press.

Dawkins, R. (1976). *The Selfish Gene*. Oxford: Oxford University Press.

Dawkins, R. (1998). *Unweaving the Rainbow: Science, Delusion and the Appetite for Wonder*. New York: Houghton Mifflin.

Deacon, J. (1999). South African rock art. *Evolutionary Anthropology* 8: 48–63.

Deichmann, U. (1996). *Biologists under Hitler*. Cambridge, MA: Harvard University Press, pp. 179–205.

Desmond, A. (1994). *Huxley: From Devil's Disciple to Evolution's High Priest*. New York: Perseus.

Diamond, M. (1990). Selected cross-generational sexual behavior in traditional Hawaii: A sexological ethnography. In J. R. Feierman (ed.), *Pedophilia: Biosocial Dimensions*, pp. 378–393. New York: Springer.

Drea, C. M., and Wallen, K. (1999). Low status monkeys "play dumb" when learning in mixed social groups. *Proceedings of the National Academy of Sciences* 96: 12965–12969.

Ducros, A., Ducros, J., and Joulian, F. (1998). *La Culture est-elle Naturelle?* Paris: Errance.

Durham, W. H. (1991). *Coevolution: Genes, Culture, and Human Diversity*. Stanford, CA: Stanford University Press.

Ehrenreich, B. (1999). The real truth about the female body. *Time*, March 8: 57–65.

Eibl-Eibesfeldt, I. (1994). *Wider die Mißtrauensgesellschaft*. Munich: Piper.

Epstein, R., Lanza, R. P., and Skinner, B. F. (1981). "Self-awareness" in the pigeon. *Science* 212: 695–696.

Flack, J. C., and de Waal, F. B. M. (2000). "Any animal whatever": Darwinian building blocks of morality in monkeys and apes. *Journal of Consciousness Studies* 7 (1–2): 1–29.

Fouts, R. (1997). *Next of Kin*. New York: Morrow.

Freeman, D. (1983). *Margaret Mead and Samoa: The Making and Unmaking of an Anthropological Myth.* Cambridge, MA: Harvard University Press.

French, M. (1985). *Beyond Power.* New York: Ballantine.

Freud, S. (1989 [1913]). *Totem and Taboo.* New York: Norton.

Freud, S. (1989 [1930]). *Civilization and Its Discontents.* New York: Norton.

Galef, B. G. (1982). Studies of social learning in Norway rats: A brief review. *Developmental Psychobiology* 15: 279–295.

Galef, B. G. (1990). The question of animal culture. *Human Nature* 3: 157–178.

Gallup, G. G. (1970). Self-awareness in primates. *Science* 67: 417–421.

Gallup, G. G. (1982). Self-awareness and the emergence of mind in primates. *American Journal of Primatology* 2: 237–248.

Garcia, J., Ervin, F. R., and Koelling, R. A. (1966). Learning with prolonged delay of reinforcement. *Psychonomic Science* 5: 121 122.

Ghiglieri, M. (1988). *East of the Mountains of the Moon: Chimpanzee Society in the African Rain Forest.* New York: Free Press.

Ghiselin, M. (1974). *The Economy of Nature and the Evolution of Sex.* Berkeley: University of California Press.

Gilliard, E. T. (1969). *Birds of Paradise and Bowerbirds.* London: Weidenfeld & Nicolson.

Goodall, J. (1990). *Through a Window.* Boston: Houghton Mifflin.

Goodall, J. (1992). Unusual violence in the overthrow of an alpha male chimpanzee at Gombe. In T. Nishida, W. C. McGrew, P. Marler, M. Pickford, and F. B. M. de Waal, (eds.), *Topics in Primatology: Vol. 1, Human Origins,* pp. 131-142. Tokyo: University of Tokyo Press.

Gould, S. J. (1981). *The Mismeasure of Man.* New York: Norton.

Gould, S. J. (July 2, 1999) The human difference. *The New York Times.*

Green, S. (1975). Dialects in Japanese monkeys: Vocal learning and cultural transmission of locale-specific vocal behavior? *Zeitschrift für Tierpsychologie* 38: 304–314.

Greenberg, G., and Haraway, M. M. (1998). *Comparative Psychology: A Handbook.* New York: Garland.

Guinet, C., and Bouvier, J. (1995). Development of intentional stranding hunting techniques in killer whale (*Orcinus orca*) calves at Crozet Archipelago. *Canadian Journal of Zoology* 73: 27–33.

Günther, M. M., and Boesch, C. (1993). Energetic costs of nut-cracking behaviour in wild chimpanzees. In H. Preuschoft and D. J. Chivers (eds.), *Hands of Primates*, pp. 109–129. Vienna: Springer.

Halstead, L. B. (1985). Anti-Darwinian theory in Japan. *Nature* 317: 587–589.

Harris, J. R. (1998). *The Nurture Assumption: Why Children Turn Out the Way They Do*. London: Bloomsbury.

Hebb, D. O. (1971). Comment on altruism: The comparative evidence. *Psychological Bulletin* 76: 409–410.

Henrich, J., and Boyd, R. (1998). The evolution of conformist transmission and the emergence of between-group differences. *Evolution and Human Behavior* 19: 215–241.

Heyes, C. (1995). Self-recognition in mirrors: Further reflections create a hall of mirrors. *Animal Behaviour* 50: 1533–1542.

Hildebrand, G. (1999). *Origins of Architectural Pleasure*. Berkeley: University of California Press.

Hinde, R. A. (1966). *Animal Behaviour: A Synthesis of Ethology and Comparative Psychology*. New York: McGraw-Hill.

Hinde, R. A. (1982). *Ethology: Its Nature and Relations with Other Sciences*. Glasgow: Fontana.

Hinde, R. A., and Fisher, J. (1951). Further observations on the opening of milk bottles by birds. *British Birds* 44: 393–396.

Hirata, S., Myowa, M., and Matsuzawa, T. (1998). Use of leaves as cushions to sit on wet ground by wild chimpanzees. *American Journal of Primatology* 44: 215–220.

Hobbes, T. (1991 [1651]). *Leviathan*. Cambridge: Cambridge University Press.

Hodos, W., and Campbell, C. B. (1969). *Scala Naturae*: Why there is no theory in comparative psychology. *Psychological Review* 76: 337–350.

Hollard, V. D., and Delius, J. D. (1982). Rotational invariance in visual pattern recognition in pigeons and humans. *Science* 218: 804–806.

Hrdy, S. B. (1979). Infanticide among animals: A review, classification, and examination of the implications for the reproductive strategies of females. *Ethology & Sociobiology* 1: 13–40.

Hrdy, S. B. (1999). *Mother Nature: A History of Mothers, Infants, and Natural Selection*. New York: Pantheon.

Huffman, M. A. (1996). Acquisition of innovative cultural behaviors in non-human primates: A case study of stone handling, a socially transmitted behavior in Japanese macaques. In C. M. Heyes and B. G. Galef (eds.), *Social Learning in Animals: The Roots of Culture*, pp. 267–289. San Diego, CA: Academic Press.

Huffman, M. A. (1997). Current evidence for self-medication in primates: A multi-disciplinary perspective. *Yearbook of Physical Anthropology* 40: 171–200.

Hume, D. (1985 [1739]). *A Treatise of Human Nature*. Harmondsworth, UK: Penguin.

Humphrey, N. K. (1976). The social function of intellect. In P. P. G. Bateson and R. A. Hinde (eds.), *Growing Points in Ethology*, pp. 303–321. Cambridge: Cambridge University Press.

Huxley, J. (1942). The origins of human drawing. *Nature* 142: 637.

Huxley, T. H. (1989 [1894]). *Evolution and Ethics*. Princeton, NJ: Princeton University Press.

Imanishi, K. (1952). *Man*. Tokyo: Mainichi-Shimbunsha (in Japanese).

Inoue, R., and Anderson, A. (1988). The Terrier's Way. *Nature* 332: 758.

Inoue-Nakamura, N., and Matsuzawa, T. (1997). Development of stone tool use by wild chimpanzees. *Journal of Comparative Psychology* 111: 159–173.

Itani, J. (1985). The evolution of primate social structures. *Man* 20: 593–611.

Itani, J., and Nishimura, A. (1973). The study of infrahuman culture in Japan: A review. In E. W. Menzel (ed.), *Precultural Primate Behavior*, pp. 26–50. Basel: Karger.

Jacob, F. (1998). *Of Flies, Mice, and Men*. Cambridge, MA: Harvard University Press.

Jewell, D. (July 14, 1997). Brave hearts. *People*.

Kalikow, T. J. (1980). Die ethologische Theorie von Konrad Lorenz: Erklärung und Ideologie, 1938–1943. In H. Mertens and S. Richter (eds.), *Naturwissenschaft, Technik und NS-Ideologie*, pp. 189–214. Frankfurt: Suhrkamp.

Kano, T. (1992). *The Last Ape: Pygmy Chimpanzee Behavior and Ecology*. Stanford, CA: Stanford University Press.

Kano, T. (1998). Comments on C. B. Stanford. *Current Anthropology* 39: 410–411.

Kawai, M. (1965). Newly-acquired pre-cultural behavior of the natural troop of Japanese monkeys on Koshima islet. *Primates* 6: 1–30.

Kellogg, W. N., and Kellogg, L. A. (1967 [1933]). *The Ape and the Child*. New York: Hafner.

Kennedy, J. S. (1992). *The New Anthropomorphism.* Cambridge: Cambridge University Press.

Keyes, K. (1982). *The Hundredth Monkey.* Coos Bay, OR: Vision Books.

Killen, M., and de Waal, F. B. M. (2000). The evolution and development of morality. In F. Aureli and F. B. M. de Waal (eds.), *Natural Conflict Resolution,* pp. 352–372. Berkeley, CA: University of California Press.

Köhler, W. (1925). *The Mentality of Apes.* New York: Vintage Books.

Kroeber, A. L. (1928). Sub-human cultural beginnings. *Quarterly Review of Biology* 3: 325–342.

Kroeber, A. L. (1963 [1923]). *Anthropology: Culture Patterns & Procesess.* New York: Harcourt.

Kummer, H. (1971). *Primate Societies.* Arlington Heights: Davidson.

Kummer, H. (1995). *In Quest of the Sacred Baboon.* Princeton, NJ: Princeton University Press.

Kunz, T. H., and Allgaier, A. L. (1994). Allomaternal care: Helper-assisted birth in the Rodrigues fruit bat, *Pteropus rodricensis. J. Zool., London* 232: 691–700.

Kurland, J. A. (1977). *Kin Selection in the Japanese Monkey.* Contributions to Primatology, vol. 12. Basel: Karger.

Kuroda, S. (1984). Interaction over food among pygmy chimpanzees. In R. L. Susman (ed.), *The Pygmy Chimpanzee,* pp. 301–324. New York: Plenum.

Ladygina-Kohts, N. N. (in press). *Infant Chimpanzee and Human Child* (F. B. M. de Waal, ed.). New York: Oxford University Press.

Leakey, R., and Lewin, R. (1992). *Origins Reconsidered.* New York: Doubleday.

Lefebvre, L. (1995). Culturally-transmitted feeding behaviour in primates: Evidence for accelerating learning rates. *Primates* 36: 227–239.

Lenain, T. (1997). *Monkey Painting.* London: Reaktion Books.

Levy, M. (1961). Dali, the quantum gun at Port Lligat. *The Studio* 162: 83–85.

Liessmann, K. P. (1996). *Der gute Mensch von Österreich.* Vienna: Sonderzahl.

Linton, R. (1936). *The Study of Man: An Introduction.* New York: Appleton-Century-Croft.

Lorenz, K. Z. (1962 [1952]). *King Solomon's Ring.* New York: Time.

Lorenz, K. Z. (1966 [1963]). *On Aggression.* London: Methuen.

Lorenz, K. Z. (1981). *The Foundations of Ethology.* New York: Simon & Schuster.

Lorenz, K. Z. (1985). My family and other animals. In D. A. Dewsbury (ed.
), *Leaders in the Study of Animal Behavior,* pp. 259–287. Lewisburg, PA:
Bucknell University Press.

Lumsden, C., and Wilson, E. O. (1981). *Genes, Mind, and Culture.*
Cambridge, MA: Harvard University Press.

Mann, A. (1972). Hominid and cultural origins. *Man* 7: 379–386.

Manning, A. (Feb. 10, 1996). On the origins of behaviour. *New Scientist.*

Marler, P., and Tamura, M. (1964). Culturally transmitted patterns of vocal
behavior in sparrows. *Science* 146: 1483–1486.

Marshall, A. J., Wrangham, R. W., and Arcadi, A. C. (1999). Does learning
affect the structure of vocalizations in chimpanzees? *Animal Behaviour*
58: 825–830.

Marshall, Thomas E. (1993). *The Hidden Life of Dogs.* Boston: Houghton
Mifflin.

Maslow, A. (1936). The role of dominance in the social and sexual behavior
of infra-human primates. Series of articles in the *Journal of Genetic
Psychology,* vols. 48–49.

Masson, J. M., and McCarthy, S. (1995). *When Elephants Weep: The
Emotional Lives of Animals.* New York: Delacorte.

Masters, R. (1989). *The Nature of Politics.* New Haven, CT. Yale University
Press.

Matsuzawa, T. (1994). Field experiments on use of stone tools by chim-
panzees in the wild. In R. W. Wrangham, W. C. McGrew, F. B. M. de
Waal, and P. Heltne (eds.), *Chimpanzee Cultures,* pp. 351–370.
Cambridge, MA: Harvard University Press.

Mayr, E. (1997). *This is Biology: The Science of the Living World.*
Cambridge, MA: Belknap.

McGrew, W. C. (1979). Evolutionary implications of sex differences in
chimpanzee predation and tool use. In D. A. Hamburg and E. R.
McCown (eds.), *The Great Apes,* pp. 441–463. Menlo Park, CA:
Benjamin/Cummings.

McGrew, W. C. (1992). *Chimpanzee Material Culture: Implications for
Human Evolution.* Cambridge: Cambridge University Press.

McGrew, W. C., and Tutin, C. E. G. (1978). Evidence for a social custom
in wild chimpanzees? *Man* 13: 243–251.

McGrew, W. C., Ham, R. M., White, L. J. T., Tutin, C. E. G., and
Fernandez, M. (1997). Why don't chimpanzees in Gabon crack nuts?
International Journal of Primatology 18: 335–374.

Mead, M. (1950). *Male and Female: A Study of the Sexes in a Changing World.* New York: Penguin.

Medawar, P. B. (1984). *The Limits of Science.* New York: Harper & Row.

Mencius (372–289 B. C.). *The Works of Mencius.* English transl. Gu Lu. Shanghai: Shangwu Publishing House.

Midgley, M. (1979). *Beast and Man: The Roots of Human Nature.* London: Routledge.

Miller, G. F. (2000). *The Mating Mind: How Sexual Choice Shaped the Evolution of Human Nature.* New York: Doubleday.

Mineka, S., Davidson, M., Cook, M., and Keir, R. (1984). Observational conditioning of snake fear in rhesus monkeys. *Journal of Abnormal Psychology* 93: 355–372.

Mitchell, R. W., Thompson, N. S., and Miles, H. L. (1997). *Anthropomorphism, Anecdotes, and Animals.* Albany, NY: SUNY Press.

Montagu, M. F. A. (1968). *Man and Aggression.* New York: Oxford University Press.

Moore, B. R., and Stuttard, S. (1979). Dr. Guthrie and *Felis domesticus* or: Tripping over the cat. *Science* 205: 1031–1033.

Moore, J. A. (1993). *Science as a Way of Knowing: The Foundations of Modern Biology.* Cambridge, MA: Harvard University Press.

Morgan, C. L. (1894). *An Introduction to Comparative Psychology.* London: Scott.

Morgan, C. L. (1903). *An Introduction to Comparative Psychology,* 2nd edition. London: Scott.

Morris, D. (1962). *The Biology of Art: A Study of the Picture-Making Behaviour of the Great Apes and Its Relationship to Human Art.* London: Methuen.

Morris, D. (1967). *The Naked Ape.* New York: Dell.

Morris, R., and Morris, D. (1966). *Men and Apes.* New York: McGraw-Hill.

Mulder, M. (1979). *Omgaan met Macht.* Amsterdam: Elsevier.

Myowa-Yamakoshi, M., and Matsuzawa, T. (1999). Factors influencing imitation of manipulatory actions in chimpanzees. *Journal of Comparative Psychology* 113: 128–136.

Nagel, T. (1974). What is it like to be a bat? *Philosophical Review* 83: 435–450.

Nakamichi, M., Kata, E., Kojima, Y., and Itoigawa, N. (1998). Carrying and washing of grass roots by free-ranging Japanese macaques at Katsuyama. *Folia primatologica* 69: 35–40.

Nakamura, M., McGrew, W., Marchant, L. F., and Nishida, T. (2000). Social scratch: Another custom in wild chimpanzees? *Primates* 41: 237–246.

Nimchinsky, E. A., Gilissen, E., Allman, J. M., Perl, D. P., Erwin, J. E., and Hof, P. R. (1999). A neuronal morphologic type unique to humans and great apes. *PNAS* 96: 5268–5273. Proceedings of the National Academy of Sciences.

Nishida, T. (1990). A quarter century of research in the Mahale Mountains: An overview. In T. Nishida (ed.), *The Chimpanzees of the Mahale Mountains*, pp. 3–35. Tokyo: University of Tokyo Press.

Nishida, T., and Hosaka, K. (1996). Coalition strategies among adult male chimpanzees of the Mahale Mountains, Tanzania. In W. C. McGrew, L. F. Marchant, and T. Nishida (eds.), *Great Ape Societies*, pp. 114–134. Cambridge: Cambridge University Press.

Nottebohm, G. (1880). *Mozartiana*. Wiesbaden. Breitkopf & Härtel.

Oakley, K. (1957). *Man the Tool-Maker*. Chicago: University of Chicago Press.

Ogawa, H. (1995). Recognition of social relationships in bridging behavior among Tibetan macaques. *American Journal of Primatology* 35: 305 310.

Ottoni, E. B., and Mannu, M. (in press). Semi-free ranging tufted capuchin monkeys (*Cebus apella*) spontaneously use tools to crack open nuts. *International Journal of Primatology*.

Parish, A. R. (1993). Sex and food control in the "uncommon chimpanzee": How bonobo females overcome a phylogenetic legacy of male dominance. *Ethology & Sociobiology*, 15: 157–179.

Parish, A. R., and de Waal, F. B. M. (2000). The other "closest living relative": How bonobos (*Pan paniscus*) challenge traditional assumptions about females, dominance, intra- and inter-sexual interactions, and hominid evolution. In D. LeCroy and P. Moller (eds.), *Evolutionary Perspectives on Human Reproductive Behavior. Annals of the New York Academy of Sciences* 907: 97–113.

Payne, K. (1998). *Silent Thunder: In the Presence of Elephants*. New York: Penguin.

Porter, D., and Neuringer, A. (1984). Musical discriminations by pigeons. *Journal of Experimental Psychology: Animal Behavior Processes* 10: 138–148.

Povinelli, D. J., et al. (1997). Chimpanzees recognize themselves in mirrors. *Animal Behaviour* 53: 1083–1088.

Premack, D., and Premack, A. J. (1994). Why animals have neither culture nor history. In T. Ingold (ed.), *Companion Encyclopedia of Anthropology*, pp. 350–365. London: Routledge.

Preston, S. D., and de Waal, F. B. M. (in press). The communication of emotions and the possibility of empathy in animals. In *Altruistic Love: Science, Philosophy, and Religion in Dialogue*. Oxford: Oxford University Press.

Ridley, M. (1996). *The Origins of Virtue*. London: Viking.

Roberts, M. (1996). *The Man Who Listens to Horses*. New York: Random House.

Roes, F. (1997). An interview of Richard Dawkins. *Human Ethology Bulletin* 12(1): 1–3.

Roes, F. (1998). A conversation with George C. Williams. *Natural History* 5: 10–15.

Russell, B. (1927). *Outline of Philosophy*. New York: Median.

Russon, A. E. (1996). Imitation in everyday use: Matching and rehearsal in the spontaneous imitation of rehabilitant orangutans (*Pongo pygmaeus*). In A. E. Russon, K. A. Bard, and S. T. Parker (eds.), *Reaching into Thought: The Minds of the Great Apes*, pp. 152–176. Cambridge: Cambridge University Press.

Sacks, O. (1985). *The Man Who Mistook His Wife for a Hat*. London: Picador.

Sakura, O. (1998). Similarities and varieties: A brief sketch on the reception of Darwinism and Sociobiology in Japan. *Biology & Philosophy* 13: 341–357.

Savage-Rumbaugh, S., and Lewin, R. (1994). *Kanzi: The Ape on the Brink of the Human Mind*. New York: Wiley.

VanSchaik, C. P., Deaner, R. O., and Merrill, M. Y. (1999). The conditions for tool use in primates: Implications for the evolution of material culture. *Journal of Human Evolution* 36: 719–741.

Schiller, P. H. (1951). Figural preferences in the drawings of a chimpanzee. *Journal of Comparative Psychology* 46: 101–111.

Sept, J. M., and Brooks, G. E. (1994). Reports of chimpanzee natural history, including tool-use, in 16th- and 17th-century Sierra Leone. *International Journal of Primatology* 15: 867–878.

Serpell, J. (1996). *In the Company of Animals: A Study of Human-Animal Relationships*. Cambridge: Cambridge University Press.

Shepard, P. (1996). *The Others: How Animals Made Us Human.* Washington, D. C. : Shearwater.

Shweder, R. A. (1991). *Thinking through Cultures.* Cambridge, MA: Harvard University Press.

Sinclair, M. (1986). Imanishi and Halstead: Intra specific competition? *Nature* 320: 580.

Small, M. F. (1998). *Our Babies, Ourselves.* New York: Anchor.

Smith, A. (1937 [1759]). *A Theory of Moral Sentiments.* New York: Modern Library.

Sober, E. (1998). Morgan's Canon. In D. D. Cummins, and C. Allen (eds.), *The Evolution of Mind*, pp. 224–242. Oxford: Oxford University Press.

Sober, E., and David Wilson, D. S. (1998). *Unto Others: The Evolution and Psychology of Unselfish Behavior.* Cambridge, MA: Harvard University Press.

Sommer, V. (1994). Infanticide among the langurs of Jodhpur. Testing the sexual selection hypothesis with a long-term record. In S. Parmigiani, and F. S. vom Saal (eds.), *Infanticide and Parental Care*, pp. 155–187. Chur: Harwood.

Stanford, C. B. (1998). The social behavior of chimpanzees and bonobos. *Current Anthropology* 39: 399–407.

Sugiyama, Y. (1967). Social organization of Hanuman langurs. In S. A. Altmann (ed.), *Social Communication among Primates*, pp. 221–253. Chicago: University of Chicago Press.

Sugiyama, Y., and Koman, J. (1979). Tool-using and -making behavior in wild chimpanzees at Bossou, Guinea. *Primates* 20: 513–524.

Tanaka, I. (1995). Matrilineal distribution of louse egg-handling techniques during grooming in free-ranging Japanese macaques. *American Journal of Physical Anthropology* 98: 197–201.

Thomas, R. K. (1998). Lloyd Morgan's canon. In G. Greenberg, and M. M. Haraway (eds.), *Comparative Psychology, A Handbook*, pp. 156–163. New York: Garland.

Thompson, R. K. R., and Contie, C. L. (1994). Further reflections on mirror usage by pigeons: Lessons from Winnie-the-Pooh and Pinocchio too. In S. T. Parker et al. (eds.), *Self-Awareness in Animals and Humans*, pp. 392–409. Cambridge: Cambridge University Press.

Thorpe, W. H. (1979). *The Origins and Rise of Ethology.* London: Praeger.

Tinbergen, T. (1963). On aims and methods of ethology. *Zeitschrift für Tierpsychologie* 20: 410–433.

Tokuda, K. (1961–62). A study of sexual behavior in the Japanese monkey. *Primates* 3(2): 1–40.

Tomasello, M. (1999). *The Cultural Origins of Human Cognition.* Cambridge, MA: Harvard University Press.

Tomasello, M., and Call, J. (1997). *Primate Cognition.* New York: Oxford University Press.

Tomasello, M., Kruger, A. C., and Ratner, H. H. (1993). Cultural learning. *Behavioral & Brain Sciences* 16: 495–552.

Tomasello, M., Savage-Rumbaugh, E. S., and Kruger, A. C. (1993). Imitative learning of actions on objects by children, chimpanzees, and enculturated chimpanzees. *Child Development* 64: 1688–1705.

Tratz, E. P., and Heck, H. (1954). Der afrikanische Anthropoide "Bonobo", eine neue Menschenaffengattung. *Säugetierkundliche Mitteilungen* 2: 97–101.

Tylor, E. B. (1871). *Primitive Culture.* London: Murray.

Vermeij, G. 1996. The touch of a shell. *Discover* 17(8): 76–81.

Vicchio, S. J. (1986). From Aristotle to Descartes: Making animals anthropomorphic. In R. J. Hoage, and L. Goldman (eds.), *Animal Intelligence: Insights into the Animal Mind,* pp. 187–207. Washington, D. C. : Smithsonian Institution Press.

Virey, J. -J. (1817). Art: Histoire naturelle. In *Nouveau dictionnaire d'histoire naturelle appliquée aux arts,* pp. 542–564. Paris: Deterville.

Visalberghi, E., and Fragaszy, D. M. (1990a). Food washing behaviour in tufted capuchins and crabeating macaques. *Animal Behaviour* 40: 829–836.

Visalberghi, E., and Fragaszy, D. M. (1990b). Do monkeys ape? In S. Parker, and K. Gibson (eds.), *"Language" and Intelligence in Monkeys and Apes: Comparative Developmental Perspectives,* pp. 247–273. Cambridge: Cambridge University Press.

Vogel, C. (1985). Evolution und moral. In H. Maier-Leibnitz (ed.). *Zeugen des Wissens,* pp. 467–507. Mainz: Hase & Koehler.

de Waal, F. B. M. (1986). The brutal elimination of a rival among captive male chimpanzees. *Ethology & Sociobiology* 7: 237–251.

de Waal, F. B. M. (1989a). *Peacemaking among Primates.* Cambridge, MA: Harvard University Press.

de Waal, F. B. M. (1989b). Behavioral contrasts between bonobo and chimpanzee. In P. Heltne, and L. A. Marquardt (eds.), *Understanding Chimpanzees,* pp. 154–175. Cambridge, MA: Harvard University Press.

de Waal, F. B. M. (1991). Complementary methods and convergent evidence in the study of primate social cognition. *Behaviour* 118: 297–320.

de Waal, F. B. M. (1995). Sex as an alternative to aggression in the bonobo. In P. Abramson, and S. Pinkerton (eds.), *Sexual Nature, Sexual Culture*, pp. 37–56. Chicago: University of Chicago Press.

de Waal, F. B. M. (1996a). *Good Natured*. Cambridge, MA: Harvard University Press.

de Waal, F. B. M. (1996b). Macaque social culture: Development and perpetuation of affiliative networks. *Journal of Comparative Psychology* 110: 147–154.

de Waal, F. B. M. (1997). *Bonobo: The Forgotten Ape*, with photographs by F. Lanting. Berkeley: University of California Press.

de Waal, F. B. M. (1998 [1982]). *Chimpanzee Politics: Power and Sex among Apes*, revised edition. Baltimore, MD: Johns Hopkins University Press.

de Waal, F. B. M. (1999). Cultural primatology comes of age. *Nature* 399: 635–636.

de Waal, F. B. M. (2000). Primates. A natural heritage of conflict resolution. *Science* 289: 586–590.

de Waal, F. B. M., and Aureli, F. (1996). Consolation, reconciliation, and a possible cognitive difference between macaque and chimpanzee. In A. E. Russon, K. A. Bard, and S. T. Parker (eds.), *Reaching into Thought: The Minds of the Great Apes*, pp. 80–110. Cambridge: Cambridge University Press.

de Waal, F. B. M., and Berger, M. L. (2000). Payment for labour in monkeys. *Nature* 404: 563.

de Waal, F. B. M., and Johanowicz, D. L. (1993). Modification of reconciliation behavior through social experience: An experiment with two macaque species. *Child Development* 64: 897–908.

de Waal, F. B. M., and Luttrell, L. M. (1988). Mechanisms of social reciprocity in three primate species: symmetrical relationship characteristics or cognition? *Ethology and Sociobiology* 9: 101–118.

de Waal, F. B. M., and Seres, M. (1997). Propagation of handclasp grooming among captive chimpanzees. *American Journal of Primatology* 43: 339–346.

Walker, A. (1998). *By the Light of My Father's Smile*. New York: Ballantine.

Watanabe, K. (1989). Fish: A new addition to the diet of Japanese macaques on Koshima Island. *Folia primatologica* 52: 124–131.

Watanabe, K. (1994) Precultural behavior of Japanese macaques: Longitudinal studies of the Koshima troops. In R. A. Gardner, A. B.

Chiarelli, B. T. Gardner, and F. X. Plooij (eds.), *The Ethological Roots of Culture*, pp. 81–94. Dordrecht: Kluwer.

Watanabe, S., and Nemoto, M. (1998). Reinforcing properties of music in Java sparrows (*Padda oryzivora*). *Behavioural Processes* 43: 211–218.

Watanabe, S., Sakamoto, J., and Wakita, M. (1995). Pigeons' discrimination of paintings by Monet and Picasso. *Journal of the Experimental Analysis of Behavior* 63: 165–174.

Watson, J. B. (1930 [1925]). *Behaviorism: Revised Edition*. Chicago: University of Chicago Press.

West, M. J., and King, A. P. (1990). Mozart's starling. *American Scientist* 78: 106–114.

Westermarck, E. (1912). *The Origin and Development of the Moral Ideas*, vol. 1. London: Macmillan.

White, L. A. (1959). *The Evolution of Culture*. New York: McGraw-Hill.

Whitehead, H. (1998). Cultural selection and genetic diversity in matrilineal whales. *Science* 282: 1708–1711.

Whiten, A. (1998). Imitation of the sequential structure of actions by chimpanzees. *Journal of Comparative Psychology* 112: 270–281.

Whiten, A., Goodall, J., McGrew, W. C., Nishida, T., Reynolds, V., Sugiyama, Y., Tutin, C. E. G., Wrangham, R. W., and Boesch, C. (1999). Cultures in chimpanzees. *Nature* 399: 682–685.

Whittemore, H., and Hebard, C. (1995). *So That Others May Live*. New York: Bantam.

Williams, G. C. (1988) Reply to comments on "Huxley's evolution and ethics in sociobiological perspective. " *Zygon* 23: 437–438.

Wilson, E. O. (1995). *Naturalist*. New York: Warner.

Wilson, E. O. (1998). *Consilience: The Unity of Knowledge*. New York: Knopf.

Wimsatt, W. C. (1999). Genes, memes, and cultural heredity. *Biology and Philosophy* 14: 279–310.

Wolf, A. P. (1995). *Sexual Attraction and Childhood Association: A Chinese Brief for Edward Westermarck*. Stanford, CA: Stanford University Press.

Wolpert, L. (1992). *The Unnatural Nature of Science*. London: Faber & Faber.

Woodward, R., and Bernstein, C. (1976). *The Final Days*. New York: Simon & Schuster.

Wrangham, R. W., and Peterson, D. (1996). *Demonic Males: Apes and the Evolution of Human Aggression*. Boston: Houghton Mifflin.

Wrangham, R. W., McGrew, W. C., de Waal, F. B. M., and Heltne, P. (1994). *Chimpanzee Cultures.* Cambridge, MA: Harvard University Press.

Wright, R. (1994). *The Moral Animal; The New Science of Evolutionary Psychology.* New York: Pantheon.

Wright, R. (Dec. 13, 1999). The accidental creationist: Why Stephen Jay Gould is bad for evolution. *The New Yorker,* pp. 56–65.

Yamakoshi, G. (1998). Dietary responses to fruit scarcity of wild chimpanzees at Bossou, Guinea: Possible implications for ecological importance of tool use. *American Journal of Physical Anthropology* 106: 283–295.

Yoshimi, K. (1998). Imanishi Kinji's biosociology as a forerunner of the semiosphere concept. *Semiotica* 120: 273–297.

Zhao, Q -K (1996). Etho-ecology of Tibetan macaques at Mount Emei, China. In J. E. Fa, and D. G. Lindburg (eds), *Evolution and Ecology of Macaque Societies,* pp. 263–289. Cambridge: Cambridge University Press.

Acknowledgments

My initial plan to introduce a wide audience to the latest findings on animal culture turned out to be pointless if I didn't at the same time explore human cultural views, which in the West are generally hostile to any blurring of the line between beast and man. Consequently, this book reflects on our relation with nature, and the usefulness of a dichotomy between culture and nature. Greatly stimulated by a sabbatical in China and Japan, I set out to juxtapose Western and Eastern thinking on these matters.

During my sabbatical, I visited numerous primate field sites, watched human behavior in unfamiliar surroundings, and enjoyed animated discussions over delicious dinners with colleagues and their students. Especially in Japan, there exists a long and rich primatological tradition, the unique significance of which is highlighted in this book.

Thanks to developments that began there, half a century ago, we are now entering a period with increasing emphasis on within-species diversity. This is a major departure from earlier obsessions with instinct and genetic adaptation. After having written this book, I am convinced more than ever that the topic of animal culture is here to stay, and will grow into one of the most exciting fields—a field with implications beyond animal behavior.

On my travels in China I was accompanied by my friend Renmei Ren with assistance from Yanjie Su and Kanghui Yan. Jiao Shao introduced me to the teachings of Mencius, and Jinhua Li impressed me deeply with the wild Tibetan macaques in Anhui Province.

In Japan, I was graciously hosted by many different colleagues, starting with Toshisada Nishida, who invited me to Kyoto University supported by a fellowship from the Japan Society for the Promotion of Science (JSPS). With Nishida—one of the first Japanese primatologists to embrace the evolutionary approach—I discussed at length his work in the Mahale Mountains, Tanzania, and the origins of Imanishiism.

Tetsuro Matsuzawa received me at the Primate Research Institute, in Inuyama, where we went over many aspects of culture, including his sushi-master comparison. Matsuzawa and his students are busy testing social trans-mission. The image that stays with me is that of a female chimpanzee sepa-rated from us by a glass wall, bringing a hefty branch to extract honey from a tiny hole in the wall—she clearly had the right idea, but still had some fine-tuning to do.

Michael Huffman, an American-born primatologist fluent in Japanese, helped me in multiple ways and arranged an enlightening encounter with Imanishi's foremost student, Jun'ichiro Itani. We also visited Huffman's stone-handling macaques on Arashiyama, which so nicely demonstrate that cultural learning doesn't require external rewards. I am particularly grateful to Kunio Watanabe, who accompanied me on my visit to distant Koshima, arranged a potato feeding of the monkeys, and brought me in contact with Satsue Mito with whom I communicated thanks to the translation skills of Satsuki Kuroki.

Other extremely helpful Japanese colleagues were Takeshi Furuichi, Chie Hashimoto, Satoshi Hirata, Shoji Itakura, Takayoshi Kano, Suehisa Kuroda, Junshiro Makino, Masayuki Nakamichi, Osamu Sakura, Yukimaru Sugiyama, Hideko Takeshita, Keiji Terao, Shigeo Uehara, Toshifumi Udono, Yoshikazu Ueno, Soshichi Uchii, Shigeru Watanabe, Juichi Yamagiwa, and Gen Yamakoshi. Toshikazu Hasegawa and Mariko Hiraiwa-Hasegawa invited me to an enjoyable weekend on Izu, from where we could see the shimmering peak of Mount Fuji, while briefing me on the reception of sociobiology in their country. Kazuhiko Hosaka served as an indispens-able travel coordinator not realizing beforehand how complex my trip to the far corners of Japan was to become.

I thank my co-worker, Filippo Aureli, for taking over the supervision of my research team of technicians and students, not to mention the primate colonies, at the Yerkes Primate Center during my absence, and Emory University for a sabbatical leave and other indispensable support. On my re-turn trip, I stayed in Helsinki, Finland, for a symposium on Edward Westermarck: I appreciate Jukka-Pekka Takala's efforts to revive interest in Westermarck's important thoughts on the evolution of morality.

In writing this book I received great help of colleagues who offered infor-mation or the use of illustrations, for which I wish to thank Otto Adang, Robert Beck, Agnes Cranach-Lorenz, Margaret la Farge, Eva-Maria Gruber, Joseph Hearst, Robert Hinde, the Imanishi family, Jun'ichiro Itani, Thomas

Kunz, Paul Lennard, Gerald Massey, Peter Markl, Desmond Morris, Amy Parish, Bruce Plante, Daniel Povinelli, Robert Pudim, Wolfgang Schleidt, Volker Sommer, Emanuela Cenami Spada, and Meredith West. I received assistance with the acquisition of illustrations from Darren Long and James Choo at the Living Links Center (www. emory. edu/LIVING_LINKS/), and Frank Kiernan helped develop and print the many photographs I took in China and Japan.

Most of this book is new, but some chapters borrow text from previous pieces that I wrote for *The Chronicle of Higher Education*, *Discover*, *Libération*, *Nature*, *The New York Times*, *Scientific American*, and *The Times Higher Education Supplement*. I am grateful for the support of Elizabeth Ziemska, my agent, and Don Fehr, my editor at Basic Books, as well as John Bergez for many constructive suggestions for improvement. Sections of the book have benefited from critical feed-back and fact-checking by Pamela Asquith, Jeanne Ferris, Harold Gouzoules, Mariko Hasegawa, Michael Huffman, Suehisa Kuroda, Satsuki Kuroki, Jinhua Li, Tetsuro Matsuzawa, William McGrew, Toshisada Nishida, Holger Preuschoft, Renmei Ren, Carel van Schaik, Osamu Sakura, Yukimaru Sugiyama, Soshichi Uchii, Kunio Watanabe, Shigeru Watanabe, Meredith West, and Andrew Whiten.

Finally, I thank my wife, Catherine Marin, for comments on all bits and pieces of text on the day of production. Together, we are living proof that cultural diversity, which makes for interesting collisions between routines and values, is easily bridged by mutual love.

Index

Culture
    and acquiring knowledge and
        habits from others, 6
    and anthropologists, 214–215
    as behavioral transmission, 214,
        237
    and biologist's definition of
        without regard for how it is
        accomplished, 215–216
    can other animals besides
        humans have, 213–238
    common ground or shared
        humanity of, 275–277, 362
    definition of, 5–6, 25–26, 31,
        177, 215, 236–238
    as distinguished of man from
        nature, 360
    and food learning, 17
    and freedom, 236
    how transmission of takes place,
        215
    and human's taking further than
        animals, 29–30
    and idea of only humans as
        having made progress to, 28
    and learning of predator or
        enemy image, 15–17
    as non-genetic spreading of
        habits and information, 30–31
    opposition to idea of in animals,
        177–178, 235–236
    as part of human nature, 8–9,
        365(n4)
    propagation of, 378(n153)
    relative nature of, 213–214
    as transmission of habits and
        information by social means,
        177
    versus nature, 5–10
    and Westermarck sexual aversion
        theory and nature, 341, 343,
        385(n250)
    as "what makes us human," 215
    widespread in nature, 177
    and Wilson's integration of via
        biology, 28–29
    See also Animal culture;
        Learning

Dali, Salvador, 175
Dandy (chimpanzee), 305, 306
Darwin, Charles, 81, 83, 224,
    386–386(n256)
    as a naturalist before becoming
        theorist, 187
    and human selfishness
        moderated by kindness, 339,
        352
    and moral sense, 352–353
    and sympathy, 352
Darwinism
    defense of by Halstead against
        Imanishi, 110–111, 112–113
    envy of and reception of
        sociobiology, 123–126
    and habitat segregation as
        alternative to, 121
    and Imanishi's ideas of harmony
        and opposition to
        reductionism, 119, 122
    as reflection of society that
        produced it, 120
    and Westermarck's bringing to
        social sciences, 337
Darwistotelian view, 81, 359
Davida (orangutan), 220–221
Dawkins, Richard, 346, 347,
    385(n256)

Humans (continued)
    and integration of ape and sushi
        master, 359–360
    and learning compared with
        apes, 20–21
    and macho evolutionary
        scenarios, 128, 129
    as observer/hunter, 63
    as only ones considered as
        having a mind, 360
    self-image of and animal tool
        use, 243–244
    and sense of self and social
        dominance, 309
    and taking of culture further
        than animals, 29–30
    and the term "society," 31–32
    and time for definition of in
        terms of common ground,
        362
    and unwillingness to accept
        behaviorism in relation to,
        50–51
Hume, David, 70–71
Humphrey, Nicholas, 153
Hundredth Monkey, The (Keyes),
    205–207
Hunting
    and anthropomorphism, 63–64
    and cooperation, 366(n17)
Huxley, Julian, 165
Huxley, Thomas Henry, 112,
    121–122, 339, 385–386(n256)
    and break with Darwin, 345,
        347, 348–349
    and ethics, 344–345, 348–349
Huxleyans, 346–347

Il Postino (motion picture), 319

Imanishi, Kinji, 88, 110–126
    and avoidance of strife,
        115–116
    background of, 114
    and continuity among all life
        forms, 114, 116
    and criticism of emphasis on
        instinct, 179, 194–195
    and culture as behavioral
        transmission, 214, 237
    decline of, 123
    and habitat segregation,
        115–116
    and Halstead's defense of
        Darwinism, 110–111,
        112–113
    and human behavior and
        culture, 179
    influence of, 119
    and Koshima and animal
        culture, 200, 201
    and lack of anti-
        anthropomorphism, 88
    and Nishida, 114–115
    and Ray Carpenter, 118–119,
        193
    and reconciliation of ideas with
        Darwinism, 122
    and sociobiology, 123–126
    and species-level control over
        individuals, 115
    and study of horses at Cape Toi,
        196–197
    and study of individual animals
        over a lifetime, 119
    tour of America of, 193
    and "when the time comes, every
        individual will change
        simultaneously," 115

A Note About the Author

Dr. Frans B. M. de Waal is the C. H. Candler Professor of Primate Behavior at Emory University and Director of the Living Links Center. One of the world's leading primate behavior experts, he is the author of *Chimpanzee Politics: Power and Sex Among Apes*, *Peacemaking Among Primates*, *Good Natured: The Origins of Right and Wrong in Humans and Other Animals*, and *Bonobo: The Forgotten Ape*.

A Note About the Type

Designed in 1935 by William Addison Dwiggins, Electra has been a standard book typeface since its release because of its evenness of design and high legibility. In the specimen book for Electra, Dwiggins himself points out the type's identifying characteristics: "The weighted top serifs of the straight letters of the lower case: that is a thing that occurs when you are making formal letters with a pen, writing quickly. And the flat way the curves get away from the straight stems: that is a speed product." Electra is not only a fine text face but is equally responsive when set at display sizes, realizing Dwiggins' intent when he set about the design: "...if you don't get your type warm it will be just a smooth, commonplace, third-rate piece of good machine technique—no use at all for setting down warm human ideas—just a box full of rivets.... I'd like to make it warm—so full of blood and personality that it would jump at you."

p 96 "backside of the mirror"
templates for learning
language acquisition Chomsky